Made Simple   HUMAN ANATOMY

This new instructive series
has been created
primarily for self-education
but can equally well
be used as
an aid to group study.
However complex the subject,
the reader is taken
step by step,
clearly and methodically,
through the course. Each volume
has been prepared by experts,
using throughout the
Made Simple technique of teaching.
Consequently the gaining
of knowledge now becomes
an experience to be enjoyed.

Accounting	Human Anatomy
Advanced Algebra & Calculus	Intermediate Algebra & Analytic Geometry
Anthropology	Italian
Art Appreciation	Latin
Art of Speaking	Law
Biology	Management
Book-keeping	Mathematics
Chemistry	New Mathematics
Commerce	Organic Chemistry
Economics	Philosophy
Electricity	Physics
Electronic Computers	Psychology
Electronics	Russian
English	Salesmanship
French	Spanish
Geology	Statistics
German	Typing

HUMAN ANATOMY Made Simple

I. MacKay Murray, M.D.

Advisory editor
Hilary Sandall, M.A., B.M., B.Ch.

Made Simple Books
W. H. ALLEN   London

Made and printed in Great Britain
by Butler & Tanner Ltd., Frome, Somerset
for the publishers W. H. Allen & Company Ltd.,
Essex Street, London WC2R 3JG

ISBN 0 491 00237 8 Casebound
ISBN 0 491 00247 5 Paperbound

Foreword

Anatomy is the description of a structure in terms of the parts that make it up. Human Anatomy is the study of the structures that make up the human body.

Human Anatomy Made Simple is designed for those who are making their first detailed study of the subject—nurses, physiotherapists, radiographers and anyone who has the care of sick and injured people. The book can be used in conjunction with a formal course or it can be used alone by readers who do not have access to models or to a body for dissection. The facts are presented in logical sequence and each new term is explained when it is introduced. The illustrations, an important, integral part of the text, are not intended to be *exact* replicas of sections of the body; most of them are simplified diagrams which convey the essential features of a structure in the clearest possible way.

Parts of the body must be visualized in three dimensions: in some instances the drawings provide a clear explanation, in others there are references to 'living anatomy' which enable you to study your own body as a guide to understanding points of structure. You can take your pulse and feel the artery pulsate as the heart pumps blood; you can use a muscle and observe the contraction and the movement produced.

This book provides a valuable introduction or revision course for all examinations in Human Anatomy. It can also be used as a companion reader for the student taking A-level G.C.E. or any other examination of comparable standard. Although the emphasis is on the medical aspects of anatomy, the book may also be useful to artists who require a basic knowledge of body structure.

HILARY SANDALL

Table of Contents

CHAPTER ONE

GENERAL TERMINOLOGY

In describing the human body we always assume that the subject is standing in the **anatomical position** as shown in Fig. 1. The subject stands erect, facing the observer, with toes pointing forward, eyes looking forward, arms by the side of the body with the palms of the hands turned forward.

The Planes of the Body

It has also been found useful for descriptive purposes to divide the body by imaginary planes. The **median plane** (shown as A in Fig. 1)

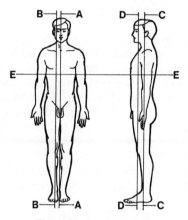

Fig. 1. Planes of the body

divides the body into symmetrical right and left halves. Any vertical plane parallel to the median plane is called a **sagittal plane** (B). In other words, the body may be divided by any number of sagittal planes and the median plane is the one in the middle. The **frontal plane** (C) is any vertical plane at right-angles to the median plane. It is also called a **coronal** plane. The plane D in Fig. 1 is also a frontal plane. The **transverse plane** (E) is any horizontal plane drawn through the body, and is of course at right-angles to both the median and frontal planes.

1

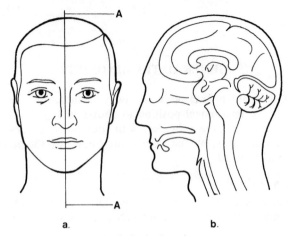

Fig. 2. Sagittal section

Some of the diagrams that appear throughout this book are described as showing a **sagittal section** or a **horizontal section.** This means that the part of the body being discussed has been cut along either the sagittal plane or the horizontal plane and you are looking at the cut surface. In a sagittal section of the head, for example, the site of the cut to be made in the sagittal plane is shown in Fig. 2a at A. A drawing is then made of either of the cut surfaces, as in Fig. 2b.

Relative Position of One Structure to Another

If we were to describe the relation of the heart to the stomach in Fig. 3a, we would say that the heart (A) is **superior** to the stomach (B).

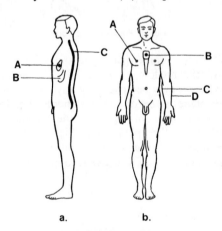

Fig. 3. Relative positions

Superior means nearer to the head. We could also say that the stomach is **inferior** to the heart. Inferior means nearer to the feet. In Fig. 3*a*, the heart (A) is **anterior** to the vertebral spine (C). Anterior means nearer to the front of the body (sometimes the word **ventral** is used instead of anterior). The vertebral spine lies **posterior** to the heart. Posterior means nearer the back (sometimes the word **dorsal** is used instead of posterior).

The term **medial** means nearer to the median plane, and **lateral** means farther from the median plane. Several medio-lateral relationships are shown in Fig. 3*b*. The nipple (A) lies lateral to the sternum (breastbone, B). However, the nipple lies medial to the arm. In the anatomical position, the inside of the upper limb (C) is called its medial side because it is the side nearest the medial plane. The side farthest from the medial plane is called the lateral side (D).

The same terms are used to refer to the limbs. For example, in the anatomical position the arms are held so that the palms are facing forward: the palmar surface of the hand is then said to be ventral or anterior. The opposite surface, i.e. the back of the hand and arm, is called the posterior or dorsal surface. Two other terms frequently used in describing structures in the limbs are **proximal** and **distal**. The reference point is the root or attachment of the limb to the trunk. The elbow is proximal to the wrist, i.e. is nearer to the root of the arm. The fingers are distal to the wrist, i.e. are farther from the root. Similarly, in the leg, the knee is distal to the thigh and proximal to the ankle.

Superficial and **deep** are used to describe the position of structures with respect to the skin. Superficial means nearer to the skin, while deep means farther from it. If we say that a nerve lies deep to a muscle, we mean that the muscle lies between the nerve and the skin. Similarly, the muscle can be said to lie deep to a layer of fat which separates it from the skin.

The terms **external** and **internal** are used in the description of the walls of cavities such as the abdomen, or a hollow organ such as the bladder. We refer to the external and internal surfaces of the bladder, for example.

CHAPTER TWO

BONES, JOINTS AND MUSCLES

Introduction

The skeleton consists of a large number of bones and pieces of cartilage, connected by joints. It provides a rigid framework for the body and makes up about 15 per cent of the total weight of the body. The flat bones of the skull give protection to the brain and are joined together in such a way that they cannot move. The bones of the thorax surround the cavity of the chest and give protection to the heart and lungs. However, because breathing involves movement, the rib cage has more flexible joints which respond to the pull of muscles between and outside the ribs.

The bones in the limbs provide rigid levers which are held together by more flexible connective tissue. Muscles stretch across these joints in such a way that the bones may be pulled into different positions when the muscles contract. Such muscles are called **voluntary**, because we can control their action. They make up more than 40 per cent of the total weight of the body.

BONES

Bones consist essentially of tissue fibres which are impregnated with calcium salts. If a bone is soaked in dilute acid, the calcium salts dissolve out, leaving only the tissue fibres. The 'bone' still has the same shape, but is now soft and can be bent easily. If a bone is heated to a very high temperature, the tissue fibres are destroyed, leaving only the calcium salts. Again the 'bone' has the same shape, but is now very brittle. Children's bones contain a higher proportion of fibres so are more likely to bend or crack, rather than break, from an external force. The bones of elderly people have lost both tissue fibres and calcium salts and so are much more fragile.

Formation of Bone

The length of time between conception and birth is called the gestation period, and during this period the body develops rapidly. At first the skeleton consists of pieces of firm elastic tissue called **cartilage**. Although strong, cartilage is not rigid enough to support the weight of the body. In the seventh week of the gestation period, bone starts to develop in the cartilage. This process, called **ossification**, begins in the middle of the shafts of the long bones. The development of the **humerus** (arm bone) is shown in Fig. 4 and can be taken as a typical example. The shape of the original cartilage resembles in miniature the shape of the adult bone. The middle part (D) is called the **diaphysis** and is also referred to as the shaft. The cartilaginous ends (E) are enlarged and after birth will form the **epiphyses** (singular: **epiphysis**).

Bone-forming cells called **osteoblasts** first appear in the centre of the diaphysis and surround themselves with thin layers of calcium salts. This is called the **primary ossification centre** (labelled A in Fig. 4 *b*) because it is the site where cartilage is first transformed into bone. The process of ossification spreads from the centre in each direction

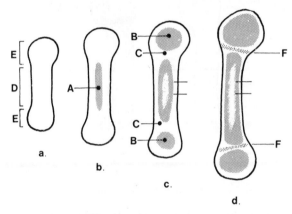

Fig. 4. Formation of bone

along the shaft. At birth, most of the diaphysis consists of bone, and bone has started to appear in the epiphyses (B in Fig. 4*c*). The epiphysis is an independent centre of bone formation, and for this reason the small mass of bone in the epiphysis is called the **secondary ossification centre.**

A disc of cartilage (C in Fig. 4*c*) separates the bone at the diaphysis from the bone of the epiphysis. This disc of cartilage is called the **epiphyseal disc** or growth disc because it is the only place where increase in length of the bone can take place. If two marks are made on the shaft of a young bone and are examined again later when the bone is older and longer, it will be found that the distance between them remains the same. Fig. 4*c* and Fig. 4*d* show how the bone grows in length at the epiphyseal discs only. When growth ceases the bony dia-physis is united with the bony epiphyses. The line of fusion is marked by a layer of dense bone (F) in Fig. 4*d*, which can often be seen on

X-ray plates of an adult bone. The age at which fusion occurs varies. The long bones of the limbs usually stop growing in late adolescence but the clavicle (collar bone) can continue to increase in length until early adult life.

Marrow

While the bones are being formed by the action of osteoblasts surrounding themselves with calcium salts, another process is taking place inside them. Gradually, as the bones become more and more solid and heavy, other cells called **osteoclasts** start to remove unnecessary calcium. These cells hollow out the insides of long bones, leaving a cavity in the middle of the diaphysis (shown at A in Fig. 5a). They

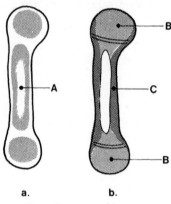

a.   b.

Fig. 5. Structure of bone

also make tiny cavities inside the ends of the long bones and inside the other bones, so that they resemble a honeycomb inside and are much lighter in weight. This type of bone is called **cancellous** bone (B in Fig. 5b). In the places where the osteoclasts have not been working the bone remains dense and hard and is called **compact** bone (C in Fig. 5b). Every bone has a layer of compact bone on the outside, although in some places this may be as thin as an eggshell. In the shafts of the long bones the outer layer of compact bone is thick and strong.

The hollowed-out spaces inside the bones are filled with **marrow**. At birth, all marrow is red and is actively producing the cells of the blood. With increasing age, the marrow becomes less active in some places; it is then composed chiefly of fat and is called yellow marrow. The cavities in the middle of the shafts of adult limb bones are filled with yellow marrow. Other bones, for example, those in the head, spine and ribs, contain active red marrow all through life. In certain diseases the marrow produces abnormal blood cells, and then it is useful to examine it. In leukaemia, for example, a specimen of marrow is obtained by puncturing the sternum (breastbone) and is examined under the microscope to determine whether abnormal white blood cells are present.

Periosteum

The skin of the bone, the **periosteum,** is shown at P in Fig. 6. Fig. 6*a* is a vertical section of a long bone, and Fig. 6*b* is a transverse section. The periosteum adheres closely to the bone, especially at the points where muscles are attached. The **tendon** of the muscle (T) is attached to the periosteum, and at this point the fibres of the periosteum penetrate the bone, anchoring the muscle firmly to the outer layer of compact bone. The ends of the bone are not covered with periosteum.

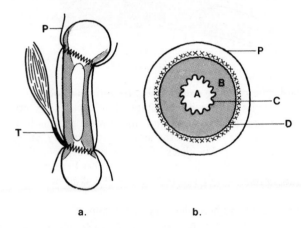

a.   b.

Fig. 6. Periosteum

Increase in the size of the bones takes place in two ways. Increase in length occurs at the epiphyseal discs, and growth in diameter occurs by the process shown in Fig. 6*b*. The marrow cavity (A) is surrounded by the compact-bone cylinder (B); the periosteum (P) surrounds the shaft. The marrow cavity becomes wider as the inner layer (C) of the compact bone is eaten away by the osteoclasts. The bone becomes wider as the cells of the periosteum lay down new bone on the outer surface (D) of the compact bone.

JOINTS

Joints are the places where two or more bones meet. Sometimes the bones are fixed together in such a way that little or no movement can occur between them. In this type of joint, the bones are bound together by fibrous tissue and so they are called either **fibrous joints** or immovable joints. Fibrous tissue is composed of threads of protein called **collagen**. These threads of collagen are very strong and they form the framework of all the tissues of the body. When they are packed closely together in dense bundles they form the fibrous connective tissue which is found in muscle tendons, joint ligaments and other structures of the body.

The bones of the skull are bound together by fibrous tissue, as shown in Fig. 7. The two bones are covered by periosteum (P), and the region where they join (J) consists of fibrous connective tissue.

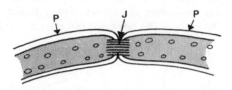

Fig. 7. Fibrous joint

A second type of joint, called a **slightly movable joint**, occurs when bones are joined together by cartilage. The cartilage is both strong and elastic: it can stretch and bend, but always returns to its former position. The ribs are joined to the sternum in this way, and so are the vertebrae of the spine. The discs of cartilage in between the vertebrae allow the spine to be flexible and also act as shock absorbers, preventing the transmission of jolts to the base of the skull during walking or running.

The third type of joint, called a **synovial joint**, allows free movement between the bones. Synovial joints are found throughout the human body: the tiny ossicles of the ear (see pages 159–60) are joined in the same way as the knee or the elbow. A diagram of a synovial joint is shown in Fig. 8. The term **articulation**, meaning joint, is frequently used.

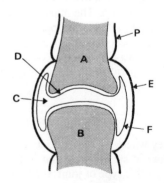

Fig. 8. Synovial joint

The ends of the bones in contact with each other are called the **articular surfaces** of the bones, i.e. the surfaces which are involved in the joint. These articular surfaces are covered with a layer of **hyaline cartilage** (labelled D in Fig. 8) which is very smooth and shiny. The articular surfaces fit closely together and are kept in position by the **fibrous capsule** (E) of the joint. This fibrous capsule forms a sleeve all round the joint and is firmly fastened to the bones where the hyaline cartilage ends and the periosteum begins. The capsule is thickened in places into **ligaments** (bands of fibrous tissue) which give additional strength to the larger joints.

The internal surface of the capsule is lined with **synovial membrane** (F), a very thin, shiny sheet of connective tissue which produces a lubricating fluid for the **joint cavity** (C). The synovial fluid is a very efficient lubricant, and synovial joints are almost completely free from friction. The amount of fluid produced depends on the physical activity of the joint. The stiffness and grating of joints observed after periods of inactivity are partly the result of a deficiency of synovial fluid. In addition to lubrication, the synovial fluid also provides nutrition for the hyaline membrane and carries away waste products.

Hyaline cartilage seems to need intermittent pressure to remain healthy. It is thicker in the leg joints where there is more weight on it. It becomes extremely thin in people confined to bed for a long time. With increasing age, small cracks and areas of softening appear, and some small areas of bone may become completely bare of cartilage.

MUSCLES

Muscle tissue consists of cells which are able to contract and thus exert a pull on whatever the muscle is attached to. There are three types of muscle in the human body: skeletal, visceral and cardiac; each is concerned with a different type of movement.

Skeletal Muscle

Skeletal muscle is also called **voluntary** muscle because we can control its action. It is attached to the skeleton, either directly or indirectly, and usually consists of a fleshy part or belly and a cord of fibrous tissue or tendon. The fleshy part is attached to one bone, while the tendon passes over a joint to become firmly attached to an adjoining bone. When the fleshy part of the muscle contracts it pulls on the tendon and so produces movement at the joint. The tendon itself does not change in length.

Feel the 'belly' of the biceps of your arm as you bend your elbow. Now try to lift the edge of a table with your palm and feel the biceps. It should be firmer, for it is contracting more strongly. Follow the belly towards the elbow and feel its cord-like tendon. When you feel a muscle, always make it perform work so that its strongly contracted belly may be felt more easily.

Each muscle is an individual organ with its own arteries, veins and nerves. Each one is enclosed in a tough cover of fibrous tissue. Some muscles are spindle-shaped like the biceps, some are long narrow strips like the sartorious, and some are flat sheets like the diaphragm.

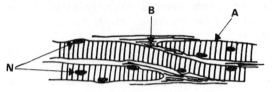

Fig. 9. Skeletal muscle

However they differ in outer appearance, all voluntary muscles are composed of the same sort of muscle fibres called **striated muscle cells.** These are shown in Fig. 9. Each cell is composed of threads of protein which are marked by alternate light and dark bands called striations. There are numerous **nuclei** (N) in each cell and connective tissue (B) between the cells (A).

A transverse section through a striated muscle (Fig. 10*a*) shows a group of cells (A) surrounded by fibrous tissue (B) to form a bundle.

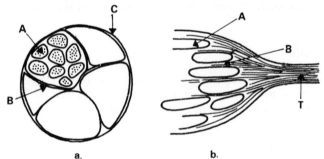

Fig. 10. Section of skeletal muscle

Many bundles, surrounded by the outer fibrous covering (C), also called the **fascia**, form the fleshy belly of the muscle. Fig. 10*b* shows how the tendon relates to the belly of the muscle. The fibrous connective tissue (B) between the muscle bundles (A) is continuous with the tendon (T).

The growth of muscles is due to an increase in size of each muscle cell. The child is born with the full number of cells, which each get longer and fatter as he grows. Exercise is responsible for growth of the muscles. On the other hand, disuse leads to shrinking in size. If an arm is kept in plaster for a few weeks, it is seen to be thinner when the plaster is removed.

Each skeletal muscle is served by a nerve. The nerve (A) serving the muscle (B) is shown in Fig. 11. A terminal branch of the nerve goes to each muscle cell and ends in a **neuromuscular plate** (C) through which the stimulus to contract is passed.

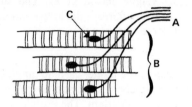

Fig. 11. Nerve supply to muscle

Normally a few cells in each muscle are being stimulated even when the muscle is at rest. This maintains 'tone', which is a state of partial contraction of the muscle. When many fibres are stimulated together, they produce either shortening of the whole muscle or an increase of tension between the two ends. One end of the muscle must remain fixed, or immobile, for it to produce movement at a joint.

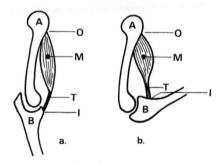

Fig. 12. Contraction of muscle

Fig. 12 shows what happens during simple flexion of a joint. The muscle M is attached to bone A at the origin O, and its tendon T passes over the joint to be inserted into bone B at I. In Fig. 12*a* the muscle is relaxed and in Fig. 12*b* it has contracted or become both shorter and fatter. The length of the tendon T remains the same. The **origin** of the muscle on bone A has not moved, but the **insertion** on bone B has moved because of the contraction. The result of the contraction has been the flexion of the joint between the two bones, or in other words, the function of the muscle is to bend, or flex, bone B relative to bone A.

Visceral Muscle

Visceral muscle is also called **involuntary** muscle because we cannot control its action. It is found in the walls of blood vessels and in the walls of the alimentary canal. The muscle cells are very small and so fine that they do not show any striations under an ordinary microscope. They are bound together in sheets by areolar connective tissue which contains yellow elastic fibres. The sheet *as a whole* is served by a network of nerve fibres—the individual muscle cells are not served by individual nerve fibres.

Cardiac Muscle

This type of muscle, as the name 'cardiac' implies, is found only in the heart. The fibres are finely striated and are joined together in a network, sometimes called a **syncytium**. The muscle contracts and relaxes rhythmically even without nervous stimulation, but the rhythm is normally controlled by nervous impulses which arrive at the sinu-atrial node and are relayed through the rest of the muscle network from cell to cell.

CHAPTER THREE

THE NERVOUS SYSTEM

Introduction

The nervous system is made up of specialized cells, called **neurones,** which receive and transmit electrical impulses. The central nervous system consists of the brain and spinal cord, while the peripheral nervous system consists of bundles of neurones which connect the central nervous system to all parts of the body. These bundles of neurones are called 'nerves' and they are distributed all over the body.

The **sensory nerves** are concerned with the transmission of electrical impulses to the central nervous system. The impulses are generated when the receptors, which are situated at the nerve-endings, respond to various stimuli. The sensory nerve-endings in the skin, for example, are sensitive to touch, pain, heat and cold.

The **motor nerves** are concerned with muscular activity or with the activity of the various glands of the body. Impulses from the central nervous system pass down the motor nerves to muscles or glands. Impulses pass down the sciatic nerve, for example, to stimulate the contraction of the gastrocnemius muscle in the back of the leg.

The activity of the neurones is to transmit electrical impulses, which are the same in all nerves, whether a sensory nerve responding to a painful stimulus, or a motor nerve stimulating a muscle to contract.

The central nervous system is an incredibly complicated pattern of neurones. They provide a system of pathways to sort out all the diverse impulses constantly arriving through the sensory nerves, and also to co-ordinate the pattern of muscular activity. The central nervous system must also provide in some way for the storage and retrieval of information. The average adult brain weighs almost three pounds and consists of millions of neurones which are grouped into different pathways and areas.

Neurones

The structural unit of the nervous system is the neurone (Fig. 13), which consists of a **cell body** (A) containing a **nucleus** (B) and **cytoplasm** (C). Two types of processes extend from the cell body: dendrites and axons.

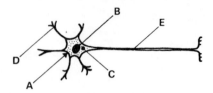

Fig. 13. Neurone

The **dendrite** (D) is usually short and appears to be a branched extension of the cytoplasm of the cell. Dendrites are probably concerned with the nutrition of the cell, and each nerve cell usually has several. The **axon** (E) may vary in length from a few millimetres to several feet; it may be very slender or slightly more substantial, but it is never more than 20 microns in diameter.

Impulses are brought to the cell body by the dendrites and are carried from the cell body by the axon. The speed of the impulse along an axon is related to its thickness. Impulses travel faster down the larger nerve fibres.

Many of the nerve fibres have a covering of fatty material called a **myelin sheath**, which gives the fibres a white appearance. They are

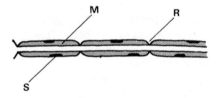

Fig. 14. Axon

called white fibres or **medullated** fibres. As shown in Fig. 14, the sheath (M) starts soon after the fibre leaves the cell body and continues almost to the branched ending of the nerve. It is interrupted at intervals along the axon cylinder, forming the **nodes of Ranvier** (R); the electrical impulses 'jump' down the axon from node to node. This type of conduction is faster than the simple spreading of the impulse down the nerve, and it is called **saltatory conduction**. Obviously, it can only occur in medullated nerve fibres, as it requires the insulating effect of the myelin sheath. The axons in peripheral nerves are also surrounded by a sheath of very delicate cells called **Schwann cells** (S), external to the myelin layer. These cells become very active when a nerve cell is damaged, and they are essential for the growth of new nerve fibres. The fibres within the central nervous system, and those of the optic nerve, have no Schwann cells and so damage to these fibres is permanent.

When an axon in a peripheral nerve is severed, the axon distal to the cut degenerates because it is separated from the cell body which provides it with nutrition. However, the nerve can grow again if the Schwann cells remain in position. The sequence of events is shown in Fig. 15. The axon distal to the cut degenerates completely and is

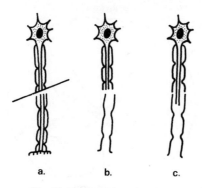

a.   b.   c.

Fig. 15. Regeneration of neurone

removed along with the myelin by phagocytes, leaving an empty tube consisting of Schwann cells (Fig. 15*b*). The cut end of the axon cylinder which is still connected to the cell body begins to grow down the hollow sheath and will finally reach the end of the nerve sheath. This process can take as long as a year. Axons will grow down any sheath, hence in repairing an injury it is important to try and join the cut ends together exactly (Fig. 15*c*) so that the original sheath is in the right position to receive the growing nerve fibre.

Synapses

When the cell body is stimulated, an electrical impulse passes along the axon to its terminal branches. At the nerve-ending the axon is touching a dendrite, a cell body, a muscle cell, or a gland, depending on its function. The point of contact between the nerve-ending and its destination is called a **synapse**. The junction between a motor nerve and a muscle fibre is called the neuromuscular junction, or the **motor end-plate**. The arrival of the electrical impulse at the end-plate causes the release of **acetylcholine** from the nerve ending. This is a chemical transmitter substance which affects the end-plate and initiates the contraction of the muscle. As soon as the impulse has passed, the acetylcholine is destroyed by an enzyme and the junction is then ready for the transmission of the next impulse.

The synapses between neurones in the central nervous system probably operate in a similar fashion, using transmitter substances to bridge the slight gap between the neurones.

THE SPINAL CORD

The **spinal cord** (Fig. 16) consists of a small central canal (A) surrounded by a mass of nerve cell bodies. Because the cells are grey when seen in a section of fresh cord, this area is called the **grey matter** (B).

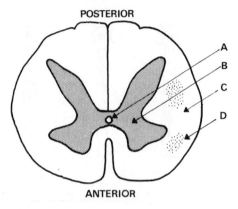

Fig. 16. Section of spinal cord

External to the grey matter is an area of **white matter,** which consists of axons in their white medullary sheaths (C). These axons are grouped into **tracts** (D) which are either ascending or descending the cord. The ascending fibres going to the brain are the **sensory tracts.** The descending fibres coming from the brain end in the grey matter of the spinal cord where they synapse with the cells of motor nerves, hence they are called the **motor tracts.**

Spinal Nerves

The anterior projection of grey matter in the spinal cord (Fig. 17) is called the **anterior horn** (A). Both anterior horns contain cell bodies of motor neurones, i.e. cells whose axons end on muscle fibres. Motor

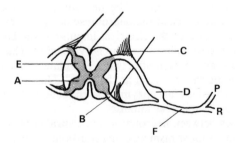

Fig. 17. Spinal nerve roots

axons leaving the cord are grouped together to form the **ventral root** (B). The **posterior root** (C) consists of the sensory axons entering the cord. The swelling or **ganglion** (D) on the dorsal root contains the cell bodies of the sensory nerves. The **posterior horns** (E) contain the cell bodies of connecting neurones of the spinal cord.

The dorsal and ventral roots unite to form a **spinal nerve** (F), which is called a mixed nerve because it contains both motor and sensory fibres. The term **spinal-cord segment** is frequently used, and is defined as a level of the cord which gives rise to a pair (i.e. right and left) of spinal nerves. There are 31 pairs of spinal nerves. Each nerve soon divides into a **posterior** branch or **ramus** (P), which supplies the muscles and skin of the back, and an **anterior ramus** (R), which supplies the lateral and anterior parts of the body. The anterior rami are larger and contain the nerves for the limbs.

Spinal Reflexes

The neurone is the basic structural unit of the nervous system. The **reflex circuit** is the basic functional unit. The neurones each have the same activity, which is to transmit electrical impulses, and they are arranged in various complex pathways. The basic pathway is the simple reflex circuit shown in Fig. 18.

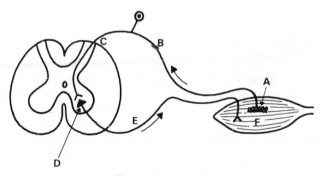

Fig. 18. Spinal reflex

When a muscle is stretched, specialized nerve endings (A) are stimulated and impulses pass up the sensory nerve (B) into the dorsal root (C) of the spinal cord. The axon of the sensory nerve acts on the anterior-horn cell (D) in the same segment of the cord. This cell sends impulses along the motor nerve (E), and the muscle fibres in the same muscle (F) contract. This constitutes the **stretch reflex**.

Stretch reflexes are found in all the muscles of the body; in other words, if any muscle is stretched it will contract promptly. This can be shown very easy in the quadriceps muscle of the thigh. Sit down, cross your legs at the knees, and give a sharp tap to the tendon just below the knee cap (A in Fig. 19). This stretches the tendon and impulses travel up the pathway already shown. The motor nerve is activated and causes a sudden automatic jerk of the leg as the muscle contracts. This reflex is protective in nature, as it prevents too much

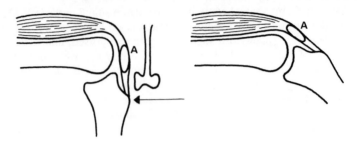

Fig. 19. Knee jerk

movement at a joint which might be injurious. The sensory stimulus of stretching triggers off a protective jerk of the muscle without conscious thought, which might take too long. Notice that it is impossible to overrule this reflex action.

AUTONOMIC NERVOUS SYSTEM

Voluntary movement is the result of the contraction of striated muscle, and the nerves already described supply striated muscle. The other types of muscle in the body (i.e. the visceral and the cardiac muscle) are called involuntary muscles, because we are not normally conscious of their activity. Involuntary muscles are supplied by the **autonomic nervous system**. Cardiac muscle is found in the heart; visceral muscle is found in the walls of blood-vessels and hollow organs such as the alimentary canal and the urinary bladder.

The muscle fibres are normally in a state of partial contraction, so they can change in two ways: they can either contract and shorten further, or they can relax and become longer. In order to bring about these two changes, the muscle fibres are supplied with two nerves, one to bring about contraction and one to bring about relaxation. The length of the muscle depends on the balance of activity between the two nerves. The autonomic nervous system is composed of the **sympathetic** and the **parasympathetic** systems. In most parts of the body, the activity of the sympathetic system is opposed by the activity of the parasympathetic system. The organs under the involuntary control of the autonomic system generally receive both sympathetic nerves and parasympathetic nerves.

Sympathetic Nervous System

The basic plan of the sympathetic nervous system is shown in Fig. 20. The spinal cord is divided into segments, and a pair of spinal nerves leaves from each segment. The segment and the corresponding nerves are named after the vertebrae of the spinal column; there are eight **cervical** (C), 12 **thoracic** (T), five **lumbar** (L) and five **sacral** (S) nerves. Sympathetic fibres leave the spinal cord between the first thoracic (T1) and the second lumbar (L2) segments.

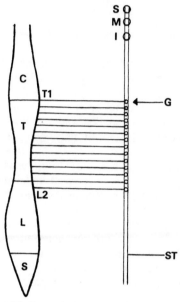

Fig. 20. Sympathetic nervous system

These fibres leave the cord with the motor nerves to voluntary muscle, i.e. in the ventral nerve roots. They run to the **sympathetic trunks** (ST) which lie on either side of the vertebral column. The sympathetic trunk consists of masses of cell bodies, **ganglia** (G), connected by nerve fibres which run vertically. The top of the sympathetic trunk lies in the neck, and the **superior cervical ganglion** (S) lies at the level of the angle of the jaw. The other two ganglia in the neck are called the **middle cervical ganglion** (M), and the **inferior cervical ganglion** (I). The disposition of the other ganglia will be described later.

The nerve fibre running from the spinal cord to a ganglion is called the **pre-ganglionic fibre** (labelled P in Fig. 21). The axon has a myelin sheath which gives the nerve fibres a white appearance. The bundle of pre-ganglionic nerve fibres is called a **white ramus**. The fibres which

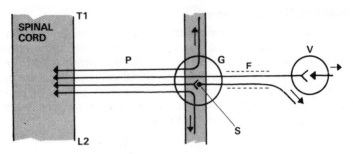

Fig. 21. Ganglion of sympathetic nervous system

start in the ganglion (G) do not have a myelin sheath, and so they are grey in colour. The bundle of **post-ganglionic fibres** (F) is therefore called the **grey ramus**. The pre-ganglionic fibre may pass through the ganglion and travel either up or down the sympathetic trunk, or it may go straight through the ganglion and travel to special ganglia on the walls of blood-vessels (V), or it may synapse on a cell body in the ganglion (S).

The post-ganglionic fibres (grey rami) usually join spinal nerves. The sympathetic fibres to the forearm and hand, for example, arrive via the three main nerves known as the median, ulnar and radial. The fibres to the head and neck, however, are found in the outer coat of the carotid arteries. The fibres to the digestive tract do not synapse in the sympathetic trunk, i.e. the pre-ganglionic fibres pass straight through the trunk as the **splanchnic nerves**, which later synapse in 'plexuses' which are found near the abdominal arteries. The post-ganglionic fibres run from these plexuses to the digestive tract.

Parasympathetic Nervous System

Parasympathetic nerve fibres originate in some areas of the brain and in the sacral segments of the spinal cord. The pre-ganglionic fibres leave the brain in the cranial nerves, III, VII, IX and X. (The twelve **cranial nerves** are usually referred to by Roman numerals.) X, the tenth nerve, or **vagus**, is the largest parasympathetic nerve in the cranial outflow. The sacral outflow is from segments 2, 3 and 4 of the sacral part of the spinal cord.

The pre-ganglionic fibres, like the sympathetic, synapse in ganglia. As a general rule, sympathetic ganglia are relatively far away from the organ they supply, i.e. the post-ganglionic fibres are relatively long. In contrast, the parasympathetic ganglia often lie in the organ that they supply. Thus the vagus nerve carries pre-ganglionic parasympathetic fibres to the heart from the brain; the ganglion and post-ganglionic fibres lie in the muscle of the heart.

Autonomic Nervous System Activity

The sympathetic nervous system is active in states of emotional excitement. It is often called the 'fight or flight' system. Its activity results in widening of the pupil, an increase in the rate and strength of the heart-beat, and increased blood flow to the voluntary muscles.

The parasympathetic system is concerned with the emptying mechanisms of the body. Its activity is necessary for the flow of saliva and gastric juice, the movement of food along the digestive tract, and the emptying of the bowel and bladder. When the parasympathetic system is active, the pupil is constricted, the blood flow to voluntary muscles is reduced, and the heart beats more slowly and less powerfully.

CHAPTER FOUR

BLOOD-VESSELS AND THE LYMPHATIC SYSTEM

Introduction

Blood moves through the body in a closed system of tubes called **blood-vessels**. The blood carries to the tissues all the things that cells require and removes from the tissues all the waste produced by the cells. This exchange can only take place through the thin delicate walls of the **capillaries**, which are very fine blood-vessels in the tissues. The heart is a pump which keeps the blood moving; the **arteries** carry the blood to the capillaries and the **veins** return it to the heart. The activity of the cells depends on the circulation of the blood: if the heart stops, the body soon dies; if the blood supply to a part is blocked, that part will soon die.

The average adult has about five litres of blood in his body. Blood is a red sticky fluid which consists of a yellow liquid called **plasma**, and a large number of cells called **corpuscles**. The red cells contain a red pigment, **haemoglobin**, which combines with oxygen. The white cells protect the body against infection. The **platelets** are concerned with the clotting of blood which prevents all the blood leaking away from small injuries to the blood-vessels. The plasma contains various substances needed by the tissues, including glucose, vitamins and salts. It also carries waste products, for example, urea and uric acid.

The functions of the blood depend on its continuous movement throughout the body. The most urgent needs of the tissues are for a fresh supply of oxygen and for the removal of carbon dioxide. The red blood cells carry oxygen to the tissues and then pick up carbon dioxide. In the lungs, the carbon dioxide leaves the red cells and diffuses into the air sacs. The red blood cells then pick up oxygen from the air sacs, and circulate back to the tissues.

Fig. 22 shows the circulation of the blood. When the heart (H) contracts, the pulmonary artery (P) carries deoxygenated blood to the lungs (L) where the red cells give up carbon dioxide and take up oxygen. The veins from the lungs return the oxygenated blood to the left side of the heart, which pumps it into the aorta (A). This is the largest artery in the body; it divides up into other arteries which supply oxygenated blood to all parts of the body. In the tissues, each artery

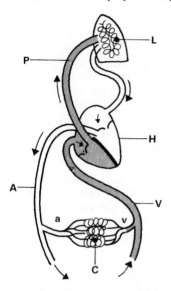

Fig. 22. Circulation of blood

splits up into finer vessels, called **arterioles** (a), which supply the capillary networks (C). Blood from the capillaries is collected into **venules** (v) and then into veins (V). The blood eventually reaches the right side of the heart via the veins, and is then returned to the lungs.

Arteries

Put two fingers on your wrist, just below the base of the thumb. You will be able to feel the radial artery pulsating as its walls suddenly expand and then recoil with every contraction of the heart. Large arteries receive the full pressure of the left ventricle and must expand enough to take up the blood which is pushed out from the heart. Then they must recoil to their former diameter as blood passes on to the capillaries. Elastic tissue in the walls of the larger arteries is responsible for the smooth expansion and recoil.

Arterioles

The amount of blood entering each arteriole with each beat of the heart is comparatively small and not under much pressure, therefore the wall does not contain much elastic tissue. Instead it is made up largely of smooth muscle. This involuntary muscle is supplied with sympathetic nerves which cause the muscle to contract, and so narrow the arteriole. When the tissues are working hard and require extra blood, the arterioles widen so that more blood can pass through them. When there is no longer any need for extra blood the arterioles contract and so reduce the supply. For example, during physical exercise the arterioles in the muscles, in the lungs and in the heart are fully open: this ensures that an adequate supply of oxygenated blood is reaching the active muscles. At the same time, the arterioles of the digestive system and urinary system are contracted, so that relatively little blood passes through these parts of the body during exercise.

Capillaries

Capillaries are thin vessels with delicate walls made up of a single layer of cells called **endothelium**. These thin walls allow fluid containing protein to leak out from the capillaries into the spaces between cells of the tissues. Because of the slightly higher pressure, fluid leaks out from the capillary at the arteriolar end and so carries nutritive substances and oxygen to the cells. At the venule end, however, the pressure is low enough to allow fluids to seep *into* the capillary, and so waste products, including carbon dioxide, are carried away from the cells into the circulation.

Veins

Blood flows from the capillaries into **venules** which join together to form veins. As the pressure of blood is very much less in the veins, their walls are much thinner than the walls of the arteries. Many veins contain **valves**, made from the projections of the lining of the walls, which ensure that the blood flows towards the heart. But not all veins have valves: they occur only where they are necessary. In the veins of the head and neck, for example, the blood flows to the heart because of gravity. Veins with valves are found most commonly in the lower limbs.

In the limbs there are two sets of veins: a **superficial** set which lie just under the skin, and a **deeper** set which generally accompany the arteries. The flow of blood is from the superficial veins to the deep veins, and valves are found at the junctions of the two sets of vessels to ensure that the flow is in this direction. If, for some reason, these valves are inefficient, blood can flow from the deep vein to the superficial vein. Because the superficial vein has very little support from the surrounding tissues, it dilates if it contains an extra load of blood. As a result, more valves become incompetent and more of the vein is dilated and twisted. This condition is called **varicose veins**.

Varicose veins are particularly common in the lower limbs because of the effect of gravity when standing upright. It may start, or become worse, in pregnancy because the foetal head presses on the iliac veins as they cross the edge of the pelvis. The varicose vein cannot efficiently remove the blood from the skin and superficial structures, and the nutrition of the skin is adversely affected by the impaired circulation. Any wound heals very slowly and ulcers may form. Methods used to treat varicose veins include compression of the superficial veins by elastic stockings, obliteration of the dilated vein by injections, or removal of the superficial veins and the closure of all the connections between the deep veins and the superficial veins.

William Harvey (1578–1657) was the first to describe the circulation of the blood correctly. He demonstrated the function of valves in veins by a simple experiment. Allow your arm to hang down until the veins on the back of your hand become distended. Choose a vein that joins another vein proximally (i.e. towards the shoulder). With two fingertips, compress the vein over its middle part. Move one fingertip proximally along the vein to its junction with the other vein, so as to press the blood out. Now remove the distal finger, (i.e. the one away from the shoulder), and you will see that the empty vein rapidly fills up to the proximal fingertip. Repeat the manœuvre, but lift the proximal fingertip instead. If a valve is present at the venous junction, the empty segment of vein remains collapsed; if there is no valve, the empty segment will fill up very slowly from above.

Collateral Circulation

In most parts of the body more than one artery supplies the same area. Not only does the capillary bed of one artery connect with that of an adjacent artery, but larger branches of one artery also directly connect with a neighbouring artery. This end-to-end joining is called an **anastomosis** and will provide an alternative route for the blood to the area if one artery is closed off. The name given to this alternative routing system is **collateral circulation.**

Many organs have an extensive collateral circulation, but there are several factors that will determine whether collateral circulation is sufficient to maintain the organ if the main artery is blocked, i.e. when there is an **arterial block.** If the vessel block is sudden, the collateral vessels may not have sufficient time to enlarge before the tissue dies. Gradual narrowing of an artery is a stimulus for the enlarging of the collateral vessels. Blockage of old arteries presents a more dangerous problem than that of young arteries. The site of the arterial block is of importance. If the block is between the heart and the point where the anastomosis takes place, the collateral vessels will not receive blood either. If an artery is surgically tied off, the site chosen is usually just beyond the area where the anastomosis between two adjacent arteries occurs.

Most organs have considerable collateral circulation, but several do not. The brain and kidney arteries do not anastomose, so they are called **end arteries.** Blockage of one artery is therefore followed by death, or **necrosis,** of the area supplied. This necrotic area is called an **infarct.** An infarct in the brain tissue is much more serious than one in the kidney. Because of the specialized function of each area in the brain, no other area can take over for the necrotic part. Since we have more kidney tissue than is needed for survival, the death of a small area is not fatal to the organism. Even a part of one kidney may be adequate for survival, which implies that the kidney has a large functional reserve.

While the small arteries in the heart muscle show some anastomosis, it is usually not sufficient to maintain the muscle if the main artery is blocked. They are regarded as functional end arteries. Whether or not the blockage of an artery to the heart muscle will be fatal is usually determined by the size of the vessel involved and hence the size of the necrotic area, i.e. the infarct.

Capillary Bypass System

In many areas of the body arterioles connect directly with venules and so bypass the capillary bed. There are several advantages in having a capillary bypass system. If all the capillaries in the voluntary muscles of the body were filled with blood, they would contain almost all the available blood in the body. When you consider the capacity of all the capillaries in the body, you can appreciate the magnitude of the surface area of the capillary bed. It is obvious that many capillaries must be shut off for varying periods of time. The blood flow through the capillary bed is controlled by the constriction of arterioles as well as by the arteriolar-venule bypass.

The major part of the body heat is lost by convection from the capillary bed in the skin. When your hand is exposed to cold, the skin becomes pale, indicating empty capillaries. Capillary bypass is a mechanism by which the body can conserve heat. The reverse takes place when the rate of heat loss from the body must be increased. During digestion, the flow of blood through the capillary bed in the walls of the alimentary tract is greatly increased, but during quiescent periods the capillary bed is bypassed. A similar pattern is followed in many other organs.

Generalized constriction of arterioles increases resistance to blood flow in the larger arteries, producing an increase in blood pressure, or **hypertension**. Surgical removal of sympathetic nerves to arterioles will decrease peripheral resistance and thus lower blood pressure.

THE LYMPHATIC SYSTEM

Although the anatomical plan of the lymphatic system was presented many years ago, its function is not yet fully understood. The lymph vessels start as blind channels in the intercellular spaces. Fluids, protein particles and inert particles such as carbon are readily taken into the lymph vessels. Small vessels join adjacent ones to form larger channels leading on to their ultimate destination at the root of the neck, where they empty into large veins. Interrupting the channels are **lymph nodes**, which in some instances seem to act as filters for materials carried in the lymph vessels.

Lymph nodes vary greatly in size. Normal lymph nodes may be felt in the groin. Swollen lymph nodes may be felt in the neck of a person with infected tonsils, or in the axilla (armpit) of a person with an infected finger. The trapping of bacteria or bacterial products by the lymph-node filters appears to be a natural defence mechanism.

In contrast, lymph vessels act as natural carriers for cancer cells. Such cells separating from a cancerous growth in the breast will travel along lymphatic vessels and when caught in a lymph node in the axilla will multiply there. The node will enlarge as a result, and its normal function will be destroyed. Succeeding cancer cells will bypass the damaged node to involve other nodes farther along the lymphatic chain. For this reason, lymph nodes receiving lymph vessels draining a cancerous area should be removed along with the cancer. Sometimes it is technically impossible to remove surgically all of the involved nodes, so the cancer frequently reappears in these nodes. It should be obvious that the early detection of cancer, followed by the surgical removal of the growth along with the accessible lymph nodes, would give the best long-term results.

Lymphocytes

Perhaps the most important function of lymph nodes is the production of lymphocytes, a type of white blood cell. For many years the lymphocyte was a cell with an unknown function. Recent evidence strongly suggests that the lymphocyte is involved in antibody formation, although its exact sequence in the chain of events is unknown. An **antibody**, a complex protein molecule, is formed in the body and aids in the destruction of all foreign material gaining entrance to the body, whether bacteria, virus, pollen, or organs transplanted from another person.

Organ transplantation is now technically possible, but destruction of the organ by antibodies frequently follows. Organs exchanged between identical twins are not destroyed by antibodies because they have developed from identical genetic material and hence are not foreign to either twin. If the lymphoid tissue is destroyed or paralysed by X-rays or chemicals, grafted organs survive for much longer periods, but the host is unable to protect himself against bacteria or viruses. When the complete story of lymphocyte function is unfolded, control of antibody production may be within reach.

CHAPTER FIVE

THE UPPER LIMB

The upper limb consists of the **arm**, which extends from the shoulder to the elbow, the **forearm**, which extends from the elbow to the wrist, and the **hand**. The joints and muscles can only be understood clearly if we first consider the various movements which are possible.

Arm Movements

In Fig. 23a the arm is shown hanging in the anatomical position, and **flexion** consists of swinging it forward and upward. Movement is

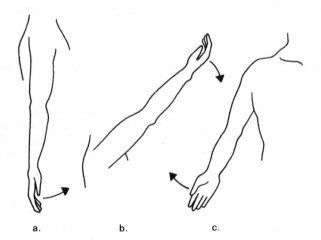

a.   b.   c.

Fig. 23. Movements of the arm

taking place between the **humerus** (arm bone) and the shoulder joint. The opposite movement is **extension**, shown in Fig. 23b. Extension starts at the end of flexion and brings the arm back to the starting position. **Abduction**, shown in Fig. 23c, is the movement of the arm away from the side of the body towards the side of the head. The word abduct means to 'lead away from'. The opposite movement, bringing the arm back to the side of the body, is called **adduction**.

28

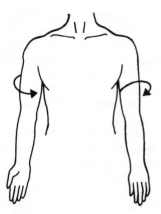

Fig. 24. Rotation of the arm

Rotation of the arm is shown in Fig. 24. Starting in the anatomical position, the right arm is turned in the direction shown by the arrow. This is medial rotation of the arm in which the lateral side is turned towards the body. The left arm in the drawing is medially rotated; the movement shown by the arrow is lateral rotation.

The shoulder, as a whole, can be **elevated**, as in shrugging, or moved in the opposite direction, which is called **depression**. It can also be moved forward, called **protrusion**, or backward, called **retraction.**

Forearm Movements

When the upper limb is in the anatomical position, the elbow joint is in extension. Movement in the direction shown by the arrow in Fig. 25*a* is flexion of the forearm. The position of flexion of the forearm is

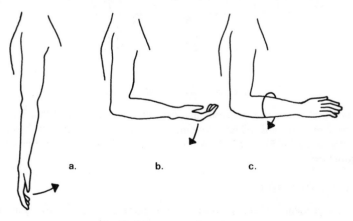

Fig. 25. Movements of the forearm

shown in Fig. 25*b*. It can be more acutely flexed and the hand can be brought up to touch the shoulder in the position of full flexion. Extension is the opposite movement, shown by the arrow in Fig. 25*b*, which brings the forearm back to the anatomical position of full extension. Fig. 25*b* shows the forearm in flexion with the palm facing upward. This position is known as **supination**. The forearm can be turned round so that the palm faces the floor; this movement is called **pronation** and is shown by the arrow in Fig. 25*c*.

Observe your hand during pronation and supination and you will see that it turns through 180° (two right-angles), the normal range of movement. Now release your elbow from the side of the body and repeat the movement; the hand can turn much more. The explanation for the greater range of movement is that the arm is free to join in when the elbow is not held to the side of the body. Rotation of the arm adds another 90° of movement to the pronation and supination of the forearm. For this reason it is important to hold the elbow to the body if you want to test the movement of the forearm only.

Hand Movements

Flexion of the hand is the movement shown by the upward arrow in Fig. 26*a*. The hand can be bent almost to a right-angle with the forearm. Extension is the movement in the opposite direction. When the

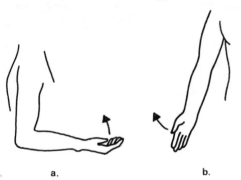

a.   b.

Fig. 26. Movements of the hand

upper limb is in the anatomical position, the hand alone can be bent away from the body. This movement is abduction of the hand, shown by the arrow in Fig. 26*b*. Movement in the opposite direction is called adduction.

THE SHOULDER

The **shoulder girdle** consists of the bones, muscles and ligaments by which the upper limb is attached to the body. The bones are the

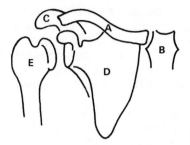

Fig. 27. Shoulder girdle

clavicle (collar bone) and the **scapula** (shoulder blade) which are shown in Fig. 27.

The clavicle (A) is a flattened bone which articulates with the **sternum** (breastbone, B) at its medial end and with the **acromial process** (C) of the scapula (D) at its lateral end. E represents the humerus (arm bone).

The scapula is shown in three different views in Fig. 28. The prominent feature on the posterior surface is the **scapula spine** (D), which separates two shallow depressions or **fossae**. The **supraspinous fossa** (A) and the **infraspinous fossa** (B) are areas from which the muscles **supraspinatus** and **infraspinatus** arise. Remember that *supra* means 'above' and *infra* means 'below'. The spine ends laterally as a blunt projection known as the **acromion** (E). The anterior surface is slightly hollowed out to form the **subscapular fossa** (C), from which the **subscapularis** muscle arises.

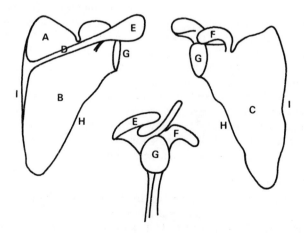

Fig. 28. Scapula

From the superior edge of the scapula, a fingerlike process points in an anterior and lateral direction; it is called the **coracoid process** (F). The **glenoid fossa** (G) is a shallow oval articular surface on the lateral aspect of the scapula. The head of the humerus fits into this fossa. The vertebral border (I) of the scapula is the edge near the vertebral column, while the axillary border (H) lies near the posterior fold of the **axilla** (armpit).

The medial end of the clavicle articulates with the sternum, forming the **sterno-clavicular joint.** The lateral end of the clavicle articulates with the acromion, forming the **acromio-clavicular joint.** (Most joints are named from the bones which form the joint.) The clavicle is also attached to the coracoid process of the scapula by the **coraco-clavicular ligament.**

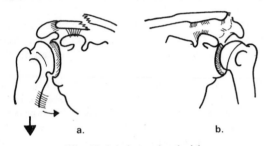

a.   b.

Fig. 29. Injuries to the clavicle

Two common injuries of the shoulder girdle are shown in Fig. 29. The commonest fracture of the clavicle, shown in Fig. 29*a* is usually the result of falling on the outstretched hand. The force travels up the humerus to the glenoid fossa. Some of this force is transmitted through the coraco-clavicular ligament to the clavicle and then to the sternum which is part of the stable framework of the trunk. If the force is excessive, the weakest point along the clavicle breaks, and the lateral fragment is drawn down by the weight of the limb and medially by spasm in the adductor muscles of the shoulder.

Fig. 29*b* shows dislocation of the acromio-clavicular joint. This injury is common among rugby football players, as it is the result of violence to the shoulder. The joint capsule and the coraco-clavicular ligaments are torn and the end of the clavicle is displaced upwards. It is easily seen and felt, and can be pushed downwards.

Rotation of the Scapula

The muscles which are attached to the scapula hold it in position, resting against the postero-lateral aspect of the rib-cage. Fig. 30 is a posterior view of the muscles which attach the scapula to the trunk. The **trapezius** (T) is shown attached to the right scapula. It is the most

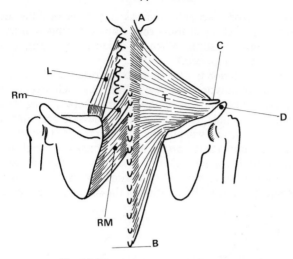

Fig. 30. Muscles attached to the scapula

superficial muscle and covers the others which are shown on the left. The origins of the trapezius extend from the **external occipital protuberance** (A) at the back of the skull, to the bony prominence of the twelfth thoracic vertebra (B). It is inserted into the clavicle (C), the acromion (D) and the spine of the scapula.

The upper part of the muscle is used to hold up the shoulder girdle (or to lift it up, as in shrugging); the middle part braces the shoulder backwards (retraction); all parts together will produce rotation of the scapula, shown in Fig. 31. This is movement around an imaginary horizontal axis through the point indicated at A. The rotated position is shown by the dotted lines and is called upward rotation. Place your

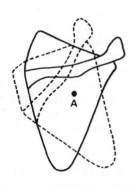

Fig. 31. Rotation of scapula

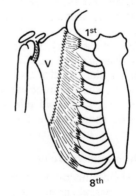

Fig. 32. Serratus anterior

hand over a person's scapula and ask him to abduct his arm. You should be able to feel the scapula rotate upwards (this is easier if your subject is a thin person). When he adducts the arm, downward rotation of the scapula occurs.

The **serratus anterior** is also involved in rotation of the scapula. It is a broad sheet of muscle, as shown in Fig. 32, which takes origin from the upper eight ribs and passes backwards between the side of the rib-cage and the scapula, to be inserted on the vertebral margin (V) of the scapula. The muscle will pull the shoulder forward (protrusion) and it also plays an important part in all movements in which the arm is raised above the horizontal. During abduction or flexion the arm has a normal range of movement of approximately 180°. One-third of the total movement is caused by rotation of the scapula. Therefore, paraly-

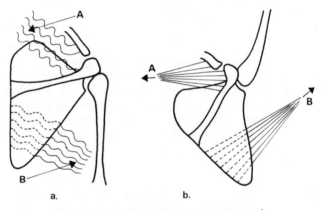

Fig. 33. Action of trapezius and serratus anterior

sis of either the trapezius or serratus anterior muscles would seriously limit flexion and abduction of the arm. The action of these muscles in rotation of the scapula is shown in Fig. 33. In Fig. 33*a* the trapezius (A) and the serratus anterior (B) are shown relaxed when the arm is at the side of the body. During upward movements (Fig. 33*b*) the two muscles contract and pull in the direction indicated by the arrows.

Three smaller muscles also hold the scapula in place against the rib-cage and are shown on the left in Fig. 30. They are the **levator scapulae** (L), the upper **rhomboid minor** (Rm) and the lower **rhomboid major** (RM). The rhomboids are used in the rotation of the scapula. The levator, as its name indicates, is used in the elevation of the shoulder.

The Gleno-humeral Joint

The gleno-humeral joint is a synovial joint of the ball-and-socket type and it permits a wide range of movement. The relatively large head of the humerus articulates with the shallow glenoid fossa. The joint capsule is fairly lax and the stability of the joint depends almost entirely on the strength of the surrounding muscles.

Certain features of the humerus and scapula, important to an understanding of the shoulder joint, are shown in Fig. 34. The articulating surface of the head of the humerus (A) is covered with articular cartilage. Distal to the head are two bony prominences called tuberosities, one larger than the other. The **greater tuberosity** (B) faces laterally, and the **lesser tuberosity** (C) is on the anterior surface of the

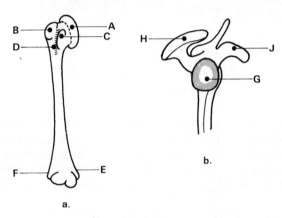

Fig. 34. Bones of the gleno-humeral joint

humerus. Between these two prominences lies a shallow groove (D) which forms a bed for one of the tendons of the biceps muscle. There are two projections at the distal end of the humerus; the **medial epicondyle** (E) is the larger, and can be felt on the medial side of the elbow. It is a reliable guide to the position of the head of the humerus, which cannot be felt, as they both point in the same direction. The **lateral epicondyle** (F) is less prominent, but can be felt on the lateral side of the elbow.

The lateral aspect of the scapula is shown in Fig. 34b. The round head of the humerus fits against the glenoid fossa (G), the acromion (H) forms a partial roof over the joint and the coracoid process (J) projects anteriorly.

MUSCLES OF THE SHOULDER AND ARM

The shoulder joint is surrounded by a loose sleeve of fibrous tissue which forms the joint capsule. In the anterior view of the joint (Fig. 35a) the capsule is shown as A. A tendon arises from the upper end of the glenoid fossa and passes inside the joint capsule to emerge at B and lie in the groove between the tuberosities of the humerus. This tendon is

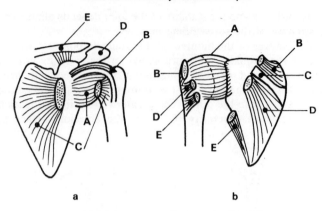

Fig. 35. Muscular cuff of the shoulder joint

the **long head** of the **biceps muscle** of the arm. The **subscapularis muscle** (C), cut in this diagram to show the joint capsule, covers the anterior aspect of the joint. It originates from the anterior surface of the scapula and is inserted into the lesser tuberosity of the humerus. The acromion (D) is superior to the joint and articulates with the clavicle (E).

In the posterior view of the joint (Fig. 35*b*) three muscles have been cut to show the capsule (A). The **supraspinatus** (B) takes origin above the spine (C), which has also been cut, and passes over the superior surface of the joint to be inserted into the greater tuberosity. The **infraspinatus** (D) and **teres minor** (E) are inserted just below the supraspinatus. The joint is thus almost completely surrounded by short muscles, which are collectively known as the muscular cuff.

Fig. 36 shows the joint from the lateral aspect with the humerus removed and the muscular cuff closely related to the joint capsule

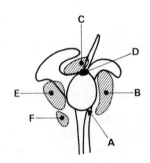

Fig. 36. Section of the muscular cuff

(A). The subscapularis (B) is anterior; the supraspinatus (C) and the long head of the biceps (D) are superior; the infraspinatus (E) and teres minor (F) are posterior. The inferior aspect of the capsule (A) is the only part not strengthened by muscles, and for this reason it is the part most commonly injured in dislocation of the shoulder. In a fall on the outstretched hand, the violent abduction of the arm may cause the head of the humerus to slip out of the glenoid fossa. It usually stretches or tears the inferior part of the joint capsule and lies below the glenoid fossa on the anterior surface of the scapula.

The Bursa

During abduction or flexion of the arm the attachment of the supra-spinatus muscle (A) comes into close contact with the overhanging acromion (B in Fig. 37). Friction between them is minimized by the

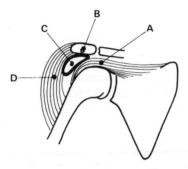

Fig. 37. Sub-acromial bursa

bursa (C) which is a fibrous bag, lined with smooth membrane and containing a small amount of viscous fluid. It lies inferior to the acromion and extends laterally to separate the **deltoid muscle** (D) from the humerus and supraspinatus. This bursa is named from its position; as it lies under the acromion it is called the **subacromial bursa**. The outer surfaces of the bursa are firmly attached to the acromion, the deltoid muscle, the muscular cuff of the shoulder joint, and the humerus. The smooth surfaces of the lining are normally in contact with each other so that, when the arm moves in flexion and abduction, they slip over one another.

Sometimes, as a result of 'everyday wear and tear', this part of the joint may become very painful. If the bursa is inflamed or swollen it may become painfully trapped between the acromion and the head of the humerus during flexion or abduction of the arm. This condition is known as **bursitis**. The tendon of the supraspinatus may also be involved in the process. Sometimes chalky deposits are found in either the tendon or the wall of the bursa.

Deltoid Muscle

The powerful deltoid muscle is shown in three separate views in Fig. 38. The anterior part of the muscle (A) arises from the clavicle and can flex the arm. The lateral part of the muscle (B) arises from the acromion and abducts the arm. The posterior part (C), arising from the

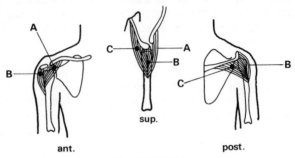

Fig. 38. Deltoid

scapula spine, can extend and abduct the arm. From its wide attachment the muscle converges to a narrow insertion about halfway down the lateral border of the humerus. This muscle gives the rounded contour to the shoulder; therefore, when it is paralysed, the shoulder is more square in outline.

Pectoralis Major

This is a flat, fan-shaped muscle on the anterior aspect of the chest wall and armpit, shown in Fig. 39*a*. One part of the muscle (A) arises from the medial half of the clavicle and is known as the **clavicular head.** The larger head (B) arises from the sternum and its lower edge (C) can

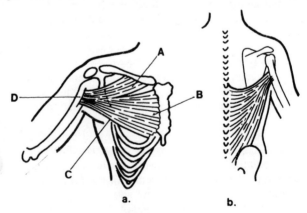

Fig. 39. Pectoralis major and latissimus dorsi

be grasped between the fingers in the anterior fold of the axilla (armpit). From its extensive attachment, the muscle converges to a narrow insertion on the lateral lip of the groove on the humerus (D). The clavicular head contracts to flex the arm, and the sternal head extends the arm. Both parts together depress the shoulder and adduct the arm. This muscle is frequently removed with a cancerous breast because the breast lies on it. Even after removal, however, flexion of the arm is still possible because the anterior part of the deltoid muscle also flexes the arm.

Latissimus Dorsi

A muscle that is well developed in strong swimmers is shown in Fig. 39*b* and is called the latissimus dorsi. It covers much of the lower part of the back and side of the trunk. It arises from a tough sheet of fibrous connective tissue called the **lumbar fascia** which is attached to the spines of the lower vertebrae. Again, it is a fan-shaped muscle which converges to a narrow tendon curving round in the posterior wall of the axilla to be inserted into the groove of the humerus, which is on its anterior aspect. It contracts to extend the arm, a movement used in swimming. It also adducts the arm, especially during extension.

Biceps

The biceps, the most prominent muscle on the anterior aspect of the arm, is shown in Fig. 40*a*. Biceps means two-headed; the long head (A) has already been described as arising inside the fibrous capsule of the gleno-humeral joint. The short head (B) arising from the cora-coid process unites with the long head to form the muscle. The tendon is inserted on to the bony prominence of the radius (C). The biceps both flexes and supinates the forearm.

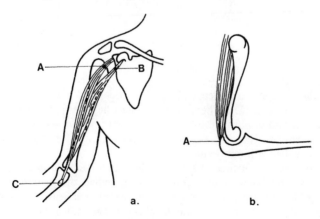

a.   b.

Fig. 40. Biceps and triceps

Triceps

On the dorsal surface of the arm there is a three-headed muscle (Fig. 40*b*). The long head arises from the scapula just below the glenoid fossa; the lateral head arises from the lateral side of the humerus and the medial head arises from the medial side of the humerus. These three heads unite to form the triceps muscle, which is inserted by a single tendon into the **olecranon process** (A) at the back of the ulna. This olecranon process is the most prominent bony point at the back of the elbow. The side view shows that contraction of the triceps causes extension at the elbow.

THE ELBOW JOINT

The elbow joint is a synovial joint and is shown in Fig. 41. At the distal end of the humerus, the rounded surface of the **capitulum** (A) articulates with the head of the radius (B). Medial to the capitulum is

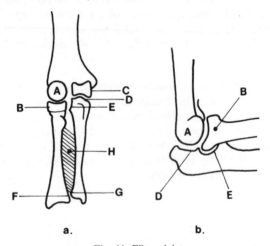

a.   b.

Fig. 41. Elbow joint

the pulley-shaped **trochlea** (C) which articulates with the **trochlear notch** of the ulna (D). These joint surfaces fit together in such a way that movement is possible in one plane alone: flexion and extension of the forearm are the only possible movements, and for this reason the elbow is called a hinge joint.

On the lateral surface of the ulna there is a rounded notch (E) into which fits the head of the radius. This is called the **proximal radio-ulnar joint**, but its synovial cavity is continuous with the synovial cavity of the elbow joint. At the distal end of the radius there is a notch (F) into which the head of the ulna (G) fits. This is called the **distal**

radio-ulnar joint. The movements of pronation and supination take place at these joints. A tough sheet of fibrous connective tissue (H) called the **interosseous membrane** joins the radius to the ulna.

Capsule of the Elbow Joint

The elbow joint is surrounded by a sleeve of fibrous connective tissue which is thicker on both the medial and lateral sides, that is, at the sides of the hinge. Another part of the capsule is thickened to form a fibrous noose around the head of the radius. This is called the **annular ligament**; it keeps the head of the radius against the ulna. If a young child is lifted by the wrist, the head of the radius may be pulled out of the annular ligament. This dislocation is called 'pulled elbow'.

Injury of the Elbow

The relative positions of the bony prominences are often important in determining whether fracture or dislocation has occurred at the elbow. Extending the forearm (Fig. 42a) and feeling the back of the

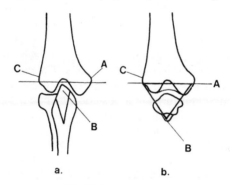

Fig. 42. Elbow triangle

elbow, you will find three bony points in the same transverse line. The prominent medial epicondyle (A), the olecranon (B) and the lateral epicondyle (C) lie in a straight line. When the forearm is flexed (Fig. 42b) these bony points form the corners of an equilateral triangle. If the bones at the elbow joint are fractured or dislocated, the three bony points will no longer show the relationships just described.

THE FOREARM

Bones of the Forearm

In Fig. 43a the forearm is shown supinated, and in Fig. 43b it is pronated. By following the radius in both diagrams you can appreciate the kind of movement that occurs. The radius during pronation turns

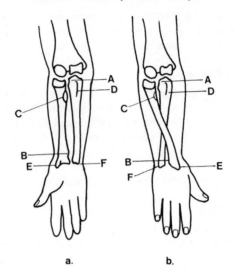

a.   b.

Fig. 43. Pronation

on its long axis and crosses the ulna. These movements occur at the proximal (A) and distal (B) radio-ulnar joints. The **radial tuberosity** (C) can easily be seen from the anterior aspect in the supinated forearm, but during pronation it turns in a posterior direction. The **ulnar tuberosity** (D) remains visible throughout because there is little, if any, movement of the ulna during pronation.

The **styloid process** of the radius (E) can easily be felt at the wrist. It extends about half an inch distal to the styloid process of the ulna (F). The force of a fall on the outstretched hand produces different effects in different age groups, but the commonest fracture is the **Colles' fracture** of the lower end of the radius (Fig. 44). The radius fractures about one inch proximal to the wrist joint; the fragment is displaced posteriorly, and the shortening which results brings the styloid processes of the radius and ulna more or less in line with each other. The shape of the wrist after this fracture has been called the 'dinnerfork deformity'.

Fig. 44. Colles' fracture

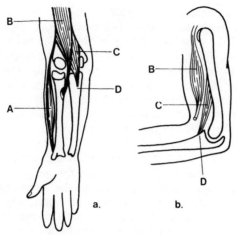

Fig. 45. Flexion of the forearm

Muscles of the Forearm

The three muscles that flex the forearm are shown in Fig. 45. Along the lateral margin of the forearm, the **brachio-radialis muscle** (A) arises from the lateral epicondyle, and is inserted about one inch from the distal end of the radius. The biceps (B) arises from the scapula as already described, and runs down the front of the arm to be inserted into the radial tuberosity. The **brachialis** (C) lies deep to the biceps. It arises from the anterior surface of the humerus and is inserted into the ulnar tuberosity (D).

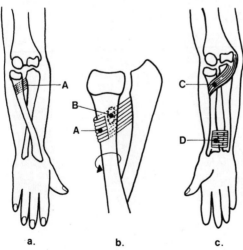

Fig. 46. Supination

Two muscles supinate the forearm. In Fig. 46*a* the forearm is pronated. The **supinator muscle** (A) arises from the ulna and is wrapped round the posterior surface of the radius to be inserted into its lateral aspect. When the supinator muscle contracts, the radius is rotated in the direction shown by the arrow in Fig. 46*b*, resulting in supination. Fig. 46*b* also shows the position of the radial tuberosity (B), which is facing in the posterior direction in the pronated radius. When the biceps contracts, its insertion in the radial tuberosity means that it rotates the radius in the direction shown by the arrow. It also flexes the forearm.

The two muscles that produce pronation are shown in Fig. 46*c*. The **pronator teres muscle** (C) arises from the medial epicondyle, as well as from the ulna, and is inserted on the lateral border of the radius. The other muscle, the **pronator quadratus** (D), arises from the ulna and is inserted into the radius. Contraction of these muscles will rotate the radius in the direction shown by the arrow, resulting in pronation.

Fig. 47 shows the muscles on the anterior aspect of the forearm. The superficial layer of muscles arises from the medial epicondyle of the humerus. The **flexor carpi radialis** (A) is inserted into the base of the second metacarpal. The **palmaris longus** (B) is not present in everybody. Its tendon, cut in this diagram, is attached to the palmar aponeurosis of the hand (see page 57). The medial muscle is the **flexor carpi ulnaris** (C) which inserts into the pisiform bone. All three muscles flex the hand at the wrist, and the flexor carpi ulnaris also adducts the hand.

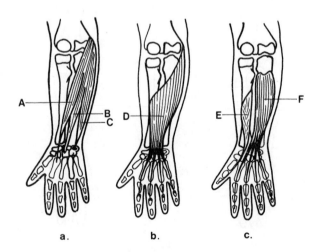

Fig. 47. Muscles of forearm (anterior)

In Fig. 47*b* the superficial layer of muscles has been removed to show the middle layer. It consists of a single broad muscle, the **flexor digitorum sublimis** (D), arising from the medial epicondyle, the ulna and the radius. Four tendons emerge from its distal end and run to the fingers. Each tendon splits just before it inserts into the base of the middle **phalanx**. (A phalanx is a finger-bone or toe-bone; the plural is phalanges.) This muscle flexes the fingers; if the fingers are prevented from flexing, it will flex the hand.

When the flexor digitorum sublimis is removed, the third layer of muscles is exposed (Fig. 47*c*). The lateral muscle (E) is the **flexor pollicis longus**. It arises from the radius and inserts into the base of the distal phalanx of the thumb. The **flexor digitorum profundus** (F) arises from the ulna and gives rise to four tendons. Each tendon inserts into the base of the terminal phalanx of a finger. As the names indicate, these muscles flex the fingers. The insertions of the flexor digitorum sublimis (Fs) and the flexor digitorum profundus (Fp) are shown in more detail in Fig. 60*a*.

The muscles on the posterior aspect of the forearm are shown in Fig. 48. The superficial muscles all arise from the lateral epicondyle, and are shown in Fig. 48*a*. The **extensor carpi radialis longus** (A) and **brevis** (B) pass distally to be inserted into the bases of the second

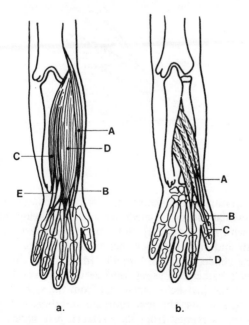

a.   b.

Fig. 48. Muscles of forearm (posterior)

and third metacarpals. Both muscles extend the hand. The **extensor carpi ulnaris** (C) is inserted into the base of the fifth metacarpal. It not only extends, but also adducts the hand. The main extensor of the fingers, the **extensor digitorum communis** (D), has four tendons which form a broad expansion at the back of the wrist, and then are inserted into the base of the middle as well as distal phalanges. A separate extensor muscle (E) for the little finger is often found.

When the superficial muscles are removed, the deep group is exposed (Fig. 48*b*). The **abductor pollicis longus** (A) is inserted into the base of the first metacarpal. The **extensor pollicis brevis** (B) is inserted into the base of the proximal phalanx. The **extensor pollicis longus** (C) passes to the base of the distal phalanx of the thumb. The **extensor indicis** (D) provides a second extensor muscle for the index finger.

BRACHIAL PLEXUS

The spinal nerves that supply the shoulder girdle and upper limb arise in the neck. They contain motor nerves which supply the muscles, and sensory nerves which carry impulses from the skin, muscle, bone and joints back to the spinal cord. These nerves form a network, or **plexus**, by dividing and recombining (Fig. 49). Although at first sight

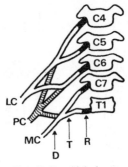

Fig. 49. Brachial plexus

the pattern appears complicated, it is fairly constant. The network, called the **brachial plexus**, is made up in the following manner.

It begins with five **roots** derived from the anterior rami of the fifth, sixth, seventh and eighth cervical nerves and the first thoracic nerve. The roots are shaded black in Fig. 49. (In the neck, the spinal nerves are named in a rather confusing way; cervical nerves 1–7 are named after the vertebrae just below their exit from the spinal cord; cervical nerve 8 emerges above the first thoracic vertebra; the first thoracic nerve, therefore, is named from the vertebra just superior to its exit, and this rule is followed for the remaining spinal nerves.) The roots link up into three **trunks**: the upper trunk from roots C5 and 6; the middle trunk from root C7; and the lower trunk from roots C8 and T1.

The trunks form six **divisions** as each trunk divides into an anterior and posterior division. The posterior divisions are shown cross-hatched in Fig. 49. These divisions link up again into three **cords**: the lateral cord (LC) formed from the anterior divisions of the upper and middle trunks; the medial cord (MC) formed from the anterior division of the lower trunk; and the posterior cord (PC) formed from the three posterior divisions.

The relationships of the brachial plexus to the bony structures in the lower neck and axilla are shown in Fig. 50. The trunks divide behind

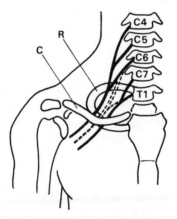

Fig. 50. Nerves to the upper limb

the clavicle, and the divisions lie posterior to the clavicle (C) but anterior to the first rib (R). The cords pass from behind the clavicle into the axilla.

Roots or trunks may be torn by any injury that stretches the plexus. Depressing the shoulder while pushing the neck to the opposite side, or violently abducting the arm, can cause injury. This can occur in accidents, in operations or in childbirth—particularly breech deliveries.

NERVE SUPPLY OF THE UPPER LIMB

The cords of the brachial plexus continue distally to form the nerves of the upper limb. The lateral and medial cords, formed from the anterior divisions, continue down the anterior aspect of the limb, as shown in Fig. 51a.

The **median nerve** (M) arises from two heads (A and B) derived from the medial and lateral cords. It supplies all the flexor muscles of the forearm except the flexor carpi ulnaris and the medial half of flexor

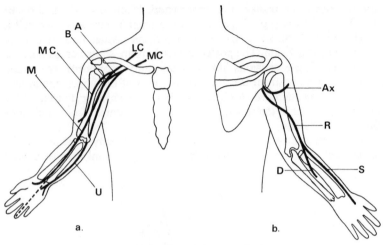

Fig. 51. Nerves of the upper limb

digitorum profundis. It supplies the muscles of the thenar eminence of the hand. Its cutaneous branches supply the skin of the hand only, and the usual distribution is shown in Fig. 52*a*.

The **ulnar nerve** (U) is a major branch of the medial cord. It passes distally to lie posterior to the medial epicondyle, where it is quite superficial. At this point, it can be felt easily (try rolling it against the bone). Because of its vulnerable position, the nerve is frequently injured when the elbow is involved in accidental damage. The medial epicondyle is called the 'funny bone' because of the sensation produced by

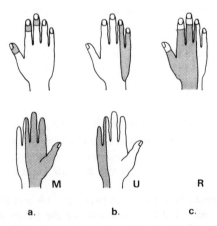

Fig. 52. Nerve supply to the skin of the hand

pressure on the ulnar nerve. In the forearm, a branch of the ulnar nerve supplies the $1\frac{1}{2}$ muscles not supplied by the median nerve. In the hand, a deep branch supplies most of the small muscles of the hand. The superficial branch supplies the skin, as illustrated in Fig. 52*b*.

The **musculo-cutaneous nerve** (MC) is the major branch of the lateral cord, which supplies the biceps and brachialis and the skin on the lateral side of the forearm. Branches from both medial and lateral cords supply the pectoralis muscle.

The **axillary nerve**, labelled Ax in Fig. 51*b*, is a major branch of the posterior cord. It passes inferior to the gleno-humeral joint, and is often injured in dislocations of that joint. The nerve winds round the posterior surface of the humerus to supply the deltoid muscle.

The **radial nerve**, labelled R in Fig. 51*b*, is the other major branch of the posterior cord. It gives branches to the triceps muscle and then lies in a groove on the posterior surface of the humerus. Fracture of the humerus in the mid-shaft may injure the radial nerve. Passing distal to the elbow, the radial nerve divides into a superficial branch and a deep branch. The superficial branch (S) passes distally to supply the skin on the dorsum of the hand, as shown in Fig. 52*c*. The deep branch (D) supplies the supinator and extensor muscles.

BLOOD SUPPLY TO THE UPPER LIMB

The blood supply to the upper limb is shown in Fig. 53. In the root of the neck, the **subclavian artery** (S) gives rise to the **internal mammary artery** (M) and the **thyro-cervical trunk** (T). It then continues as the arterial trunk to the upper limb, and at the outer border of the first rib (A) it changes its name to the **axillary artery**. The axillary

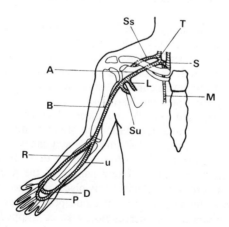

Fig. 53. Arteries of the upper limb

artery gives off several branches, which include the **lateral thoracic artery** (L) to the breast, and the **sub scapular artery** (Su) passing to the back of the scapula. As the axillary artery enters the arm, the name is again changed to the **brachial artery** (B). At the elbow, this divides into the radial (R) and ulnar (U) arteries. The radial artery is pressed against the radius at the wrist in order to 'feel the pulse'. The radial and ulnar arteries join each other in the hand to form the superficial (P) and the deep (D) palmar arches. From these arches, digital branches arise to supply the thumb and fingers.

The branches of the arteries at the top of the arm illustrate the principle of collateral circulation. The thyro-cervical trunk gives off a branch called the **suprascapular artery** (Ss) which passes down to supply the back of the scapula. This area is also supplied by the subscapular artery which leaves the axillary artery lower down. The breast is supplied by two arteries: the internal mammary and the lateral thoracic.

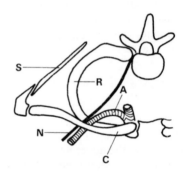

Fig. 54. Bony canal between neck and arm

Fig. 54 shows the bony canal between the neck and the arm, seen from above. The clavicle (C) is the anterior boundary, the first rib (R) is medial, and the upper border of the scapular (S) is posterior. The subclavian artery and its continuation, the axillary artery (A), are shown passing through the canal between the first rib and the clavicle, accompanied by the brachial plexus (N).

Because the rib and clavicle are close together in this area, any abnormality may interfere with either the blood or nerve supply to the upper limb. If the arm is pulled downwards and backwards, the clavicle may be pulled against the first rib so closely that the artery is compressed between them and the blood supply to the arm is cut off. A badly united fracture of the clavicle may also compress the artery. This interference with the smooth flow of blood predisposes to thrombosis and other disturbances. The nerves may also be trapped, and signs of this are numbness, tingling, and weakness of the hand.

THE HAND AND WRIST

The bones of the wrist and hand are shown in Fig. 55 from the anterior aspect. The **carpus** is composed of eight bones which constitute the wrist. They are arranged in a proximal and distal row and held together firmly by ligaments. The proximal row is made up of the **scaphoid** (S), the **lunate** (L), the **triquetral** (Tri) and the **pisiform** (P). Those of the distal row are the **trapezium** (Tr), the **trapezoid** (Tra), the **capitate** (C) and the **hamate** (H). There are five **metacarpals** (M),

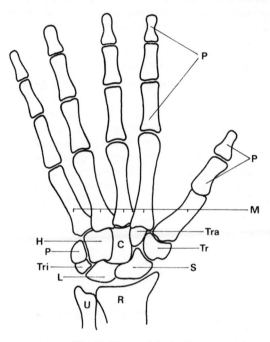

Fig. 55. Bones of the hand

the first of which is situated at the base of the thumb. The fifth lies in the ulnar side of the hand. Each metacarpal consists of a **base** which articulates with a carpal bone, a narrow tapered **shaft**, and a **head** which articulates with a **phalanx** (P). Each finger has three phalanges: proximal, middle and distal. The thumb has only two: proximal and distal.

The wrist joint is formed between the proximal row of carpal bones and the lower end of the radius. The ulna is separated from the carpal bones by an articular disc which forms part of the distal radio-ulnar joint. The proximal row of carpal bones articulates with the distal row by a transverse intercarpal joint. This joint is very mobile during flexion of the wrist. The degenerative changes of **osteoarthritis** are fairly common at the wrist because of the frequency of injuries to the joint. Movements are limited and painful in this condition.

The joint between a metacarpal and the proximal phalanx is called the **metacarpo-phalangeal joint**, a term which is sometimes abbreviated to **M–P joint.** The joints between the proximal and middle phalanges are called the **proximal inter-phalangeal joints** or **proximal I–P joints,** and those between the middle and distal phalanges are called the **distal I–P joints.**

The prominent knuckles observed when you make a fist are formed by the heads of the metacarpals. The most prominent knuckle is the head of the third metacarpal.

Movements and Muscles of the Hand

To produce flexion of the fingers, make a fist. Flexion occurs at the I–P joints and also at the M–P joints. Extension is the opposite movement, i.e. straightening of the fingers. Abduction and adduction of the fingers occur at the M–P joints. The axis of reference for these movements is a line along the middle or third finger. Spread your fingers out: this is abduction of the index, ring and little fingers, or movement away from the middle finger. The opposite movement of bringing the fingers close together is called adduction. Since the axis of reference is a fixed line, the middle finger may be abducted to either side of this line, towards the index finger or towards the ring finger.

The movements of the thumb resemble those of the fingers, but they

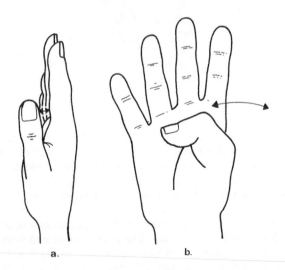

a.   b.

Fig. 56. Movements of the thumb

occur in a plane at right-angles to the fingers. Place your thumb close to the palm of your hand with thumbnail at right-angles to the finger-nails. Abduction of the thumb (Fig. 56a) is the movement away from the palm of the hand; adduction is the opposite. Flexion of the thumb (Fig. 56b) is the bending of the thumb across the palm, keeping contact with the palm during the movement. Extension is movement in the opposite direction, until the thumb is fully stretched.

The most important movement of the thumb is **opposition**. This is when the tip of the thumb touches the tip of any one of the fingers. Any pincer movement of the thumb and fingers, for example, holding a pencil in writing, requires the movement of opposition of the thumb. At least 30 per cent of the efficiency of the hand is lost if one is unable to perform this movement, which has been described as the most important single movement of the hand.

The long flexor tendons of the fingers and thumb, described previously as arising from the anterior aspect of the forearm, cross the wrist to enter the palm. As they cross the anterior, or ventral, surface of the carpal bones, they are held against the bones by a strong ligament called the **flexor retinaculum**. This prevents 'bow stringing' of the tendons during flexion of the fingers. The lateral side of the retinaculum is attached to the trapezium and the scaphoid bones. The medial side is attached to the hook of the hamate and the pisiform bones.

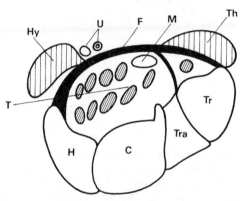

Fig. 57. Section of wrist showing carpal tunnel

A transverse section through the wrist is shown in Fig. 57. The palmar aspect of the carpal bones forms a concave surface, which is roofed over by the flexor retinaculum (F). The four bones shown are the hamate (H), the capitate (C), the trapezoid (Tra) and trapezium (Tr). They, and the other carpal bones, make up the floor and walls of the **carpal tunnel** which is roofed over by the flexor retinaculum. The long flexor tendons (T) of the fingers and thumb, and the median nerve (M) are packed into the tunnel. Any injury or disease of the walls of the tunnel may cause harmful pressure to the structures inside it. The most vulnerable is the median nerve, and damage to this nerve results in odd sensations, numbness or muscle weakness in its distribution. The ulnar nerve and artery (U) enter the palm superficial to the retinaculum. The fleshy mass at the base of the thumb (Th) also lies superficial to the retinaculum, and so does the smaller fleshy mass at the base of the little finger (Hy).

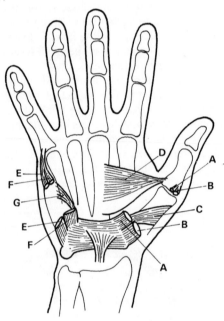

Fig. 58. Muscles of the palm of the hand (superficial)

The fleshy mass at the base of the thumb is formed by three muscles often called the **thenar muscles** (Fig. 58). Two of them, the flexor **pollicis brevis** (A) and the **abductor pollicis brevis** (B), have been partially removed in this diagram. The origins are from the lateral edge of the flexor retinaculum, and the insertions are into the base of the proximal phalanx. The third thenar muscle, the **opponens pollicis** (C), is inserted into the shaft of the first metacarpal and it also takes origin from the flexor retinaculum. A deeper muscle in the palm is the **adductor pollicis** (D), which arises from the middle metacarpal and is inserted into the proximal phalanx of the thumb. The names of these muscles are derived from their actions; see if you can feel them contract as you hold the base of your right thumb in your left hand and move it about.

The smaller fleshy mass at the base of the little finger is formed by **hypothenar muscles,** which have the same relation to the little finger as the thenar muscles have to the thumb. The **abductor digiti minimi** (E) and **flexor digiti minimi** (F) have been partially cut away to show the **opponens digiti minimi** (G) at its insertion on to the shaft of the fifth metacarpal. Obviously, the movements produced by these muscles are not of the same functional importance as those produced by the thenar muscles.

The deepest muscles of the palm are the **interossei,** shown in Fig. 59, which fill the spaces between the metacarpals. The **palmar**

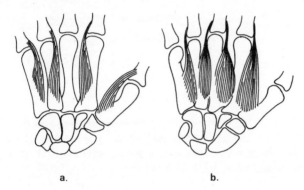

a.   b.

Fig. 59. Muscles of the palm of the hand (deep)

interossei, shown in Fig. 59a, arise from the sides of the first, second, fourth and fifth metacarpals and have long tendons which are inserted into the dorsal aspects of the proximal phalanges. Each of the four **dorsal interossei** (Fig. 59b) arises by two heads from adjacent metacarpals and is inserted into the dorsum (back) of a finger. A useful guide for remembering the functions of the interossei muscles is found in the words '*pad*' and '*dab*'. That is, the *p*almar muscles *ad*duct the fingers, and the *d*orsal muscles *ab*duct the fingers.

The action of the interossei muscles can best be understood if they are considered together with the insertion of the extensor muscles of the fingers, shown in Fig. 60. The extensor tendons (T) are inserted into the base of the proximal phalanx of each finger, and so extend the M–P joints. From this insertion the tendon spreads out distally as a wide fibrous band on the dorsal surface of the proximal and middle phalanges. This band is called the **extensor expansion** (E) and is attached to the middle phalanx and, by two lateral slips, to the distal phalanx. Into this expansion are inserted the interossei (I) and the **lumbrical muscles** (L) which arise from the flexor tendons (Fp). The interossei and lumbrical muscles flex the M–P joints and extend the I–P joints by pulling on the extensor expansion. The insertion of the

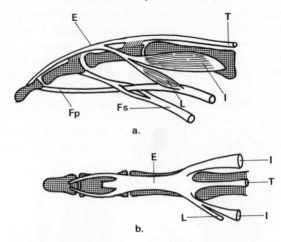

Fig. 60. Insertion of tendons in a finger

interossei on to the sides of the extensor expansion show how they are responsible for abduction and adduction. The tendon of flexor profundus (Fp) passes through that of flexor sublimis (Fs). (Fig. 59 and Fig. 60*b* should be considered together).

Nerves of the Hand

As we saw in Fig. 57, two major nerves enter the palm. The median nerve supplies the major part of the thenar muscles, while the ulnar nerve supplies the remaining intrinsic muscles of the hand. Corresponding areas of skin are supplied by these nerves, as shown in Fig. 52. Since the tendons on the dorsum of the hand all arise from muscles on the dorsal aspect of the forearm, they are supplied by the radial nerve.

Synovial Sheaths of the Hand

The long flexor tendons of the fingers pass through a confined space (the carpal tunnel) to reach the palm. Movement in such a confined space could result in considerable friction and injury. To prevent this, the tendons are covered in sheaths of synovial membrane, similar to that found in synovial joints. This facilitates smooth, gliding movements, and is shown in Fig. 61*a*. The long flexor tendons to the fingers are surrounded by a common sheath often called the **ulnar bursa** (B). This begins just proximal to the flexor retinaculum (R) and extends to the middle of the palm. The sheath of the little finger is continuous with the ulnar bursa, but the index, middle and ring fingers have separate sheaths. The long flexor of the thumb is enclosed in a separate sheath called the **radial bursa** (F). The radial and ulnar bursae usually communicate with one another deep to the flexor retinaculum.

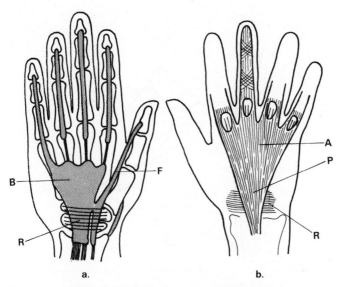

Fig. 61. Synovial sheaths and palmar aponeurosis

These synovial sheaths may become infected directly, for example, following a dog-bite, or infection may spread from a wound in a finger. Sepsis in the thumb and little finger may spread into the palm through the radial bursa and ulnar bursa, respectively. It may spread through into the forearm. Infection in the other tendon sheaths usually remains confined to the finger concerned, but may rupture through the proximal end of the sheath into the deep spaces of the palm.

Dense fascia covers the superficial aspect of the flexor tendons in the palm (Fig. 61*b*). This protects them and spreads over most of the palmar surface of the hand. The central part of the palm is covered by a dense layer called the **palmar aponeurosis** (A). It extends from the flexor retinaculum (R), where it receives the **palmaris longus tendon** (P), to the bases of the fingers where it blends with the fibrous flexor sheaths of the fingers and is attached to the sides of the proximal and middle phalanges.

THE FEMALE BREAST

The female breast is made up of 15–20 lobes of glandular tissue (Fig. 62, G) drained by **lactiferous ducts** (D). The ducts enlarge to form the **lactiferous sinuses** (S), where milk can be stored, just before they open on to the **nipple** (N). The lobes of glandular tissue are embedded in fat (F), which accounts for the smooth contour and most of the bulk of the female breast. The lobes are surrounded by fibrous tissue which

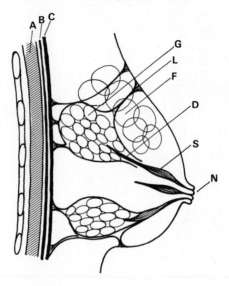

Fig. 62. Sagittal section of female breast

connects with the subcutaneous tissue and also with the fascia of the chest wall (C). These strands are called the **ligaments of Cooper** (L).

Cancer of the glandular tissue involves the fibrous tissue and produces tension of the ligaments. A late sign of cancer of the breast is dimpling of the skin resembling the skin of an orange. The breast is separated from the underlying muscle (A) by the fascia of the muscle (B) and by a layer of superficial fascia (C). The normal breast is freely movable over the underlying muscle. However, in cancer of the breast the fascia and muscle are eventually involved and another late sign is fixation of the cancerous breast to the chest wall.

The female breast overlies the second to the sixth rib; it rests on pectoralis major and serratus anterior muscles, and it just overlaps the upper part of the abdominal wall. A tail of glandular tissue usually extends into the axilla.

Blood Supply of the Breast

The blood supply comes from three sources (Fig. 63*a*). The lateral thoracic artery runs from the axillary artery (A) and supplies the lateral part of the gland. Branches from the intercostal arteries (B) pierce the chest muscles to enter the deep surface of the gland, while branches from the internal mammary artery (C) supply the medial part of the breast.

The lymphatic drainage follows the arterial supply and is of considerable importance in the spread of breast cancer. It is shown in Fig.

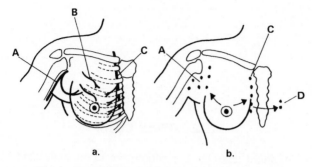

Fig. 63. Blood supply and lymphatic drainage of breast

63*b*. The major drainage is to the axilla (A). Thus, during the surgical removal of a cancerous breast, the axillary nodes are also removed. Other lymphatics follow the intercostal vessels through the pectoralis muscle. This also is removed with a cancerous breast. Lymph from the medial side of the breast drains to the internal mammary chain of nodes (C). From here it can spread across the midline to the opposite chain (D).

Areola

The circular area of pink skin surrounding the nipple is called the **areola**. During the second month of pregnancy it becomes more pigmented, i.e. darker, and the increased pigmentation is permanent.

The normal breast feels smooth when pressed with the flat of the hand against the chest wall. Any lumps, whether due to cancer or infection, can be felt easily by this method. If a woman gets into the habit of washing her breasts by rubbing them with soapy water using the flat of her hand, she will immediately discover the appearance of any lump and so will be able to ask for the early treatment which is necessary to prevent the spread of cancer.

CHAPTER SIX

THE LOWER LIMB

In the standard anatomical position, the lower limbs and feet are together and the toes point forward. The **thigh** extends from the **groin** to the knee, and the **leg** from the knee to the ankle. The **foot** is distal to the ankle. The limb has three joints: the **hip joint**, the **knee joint** and the **ankle joint**.

Thigh Movements

Medial rotation of the thigh is performed by turning the lateral side of the thigh to the midline (shown by the arrow on the left leg in Fig. 64*a*).

Lateral rotation is the reverse movement (shown by the arrow on the right leg in Fig. 64*a*). Abduction of the thigh is shown in Fig. 64*b*. Abduction of the right thigh is movement away from the midline.

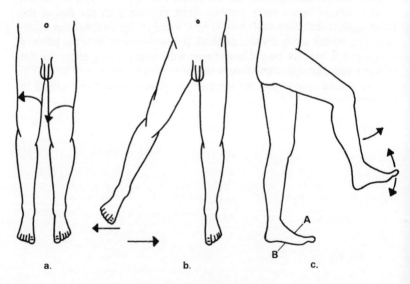

a.   b.   c.

Fig. 64. Movements of the leg

Movement in the opposite direction, towards the midline, is adduction. In Fig. 64c the left thigh and leg are extended, the right thigh and leg are both flexed, and the arrow below the knee shows the direction of extension of the leg. The hip joint is a ball-and-socket joint which permits a wide range of movements: abduction, adduction, flexion, extension, medial rotation and lateral rotation. The knee joint is a hinge joint, and so the only major movement of the knee is flexion and extension. However, as we shall see later, a small amount of rotation can also take place at this joint.

Ankle Movements

Before movements occurring at the ankle joint are discussed, the surfaces of the foot should be identified. In Fig. 64c, A marks the superior (upper) surface of the foot, which is referred to as the **dorsum** of the foot. The inferior (lower) surface (B) is called the **plantar aspect** of the foot. Both feet are shown in the neutral position. Bending the dorsum of the foot closer to the front of the leg is called **dorsiflexion**, literally 'flexion of the dorsum'. The opposite movement is called **plantar flexion** of the foot. No other movements of the foot are possible at the ankle joint.

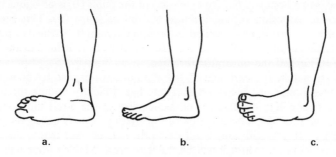

a. b. c.

Fig. 65. Movements of the foot

However, the foot can perform the movements shown in Fig. 65. The neutral position is shown at *b*. At *a*, the plantar surface is turned outward. This is called **eversion** of the foot, literally 'turning out'. At *c*, the plantar surface is turned inward, and this is called **inversion** of the foot. Eversion and inversion take place at the joints between the foot bones and *not* at the ankle.

THE PELVIC GIRDLE

The bony girdle of the lower limb is formed by the **hip bone**. Fig. 66a illustrates the external or lateral aspect of a child's hip bone. It consists of three bones joined by cartilage: the **ilium** (A), **pubis** (B)

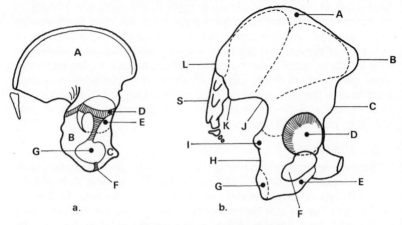

Fig. 66. Lateral aspect of the hip bone

and **ischium** (C). These bones are held together by the Y-shaped cartilage (D), which passes through the **acetabulum** (E). The acetabulum is the bony cup that receives the ball-shaped head of the **femur** (thigh bone). A barlike extension of the pubis (B) is also joined to a similar extension of the ischium (C) by cartilage (F). This bridge between the two bones is called the **conjoined rami** (branches) of pubis and ischium. Just superior to the conjoined rami is a hole in the hip bone (G) called the **obturator foramen.**

In the adult, the cartilage of the hip bone is replaced by bone and the three bones are fused, so that one has difficulty finding the lines where fusion has occurred. The same view of the adult hip bone is shown in Fig. 66*b*. The superior rim of the hip bone is called the **iliac crest** (A). If you run your hand down the side of your trunk, the first bony prominence you feel will be the iliac crest. Moving your hand in an anterior direction along the crest, you can feel the **anterior superior iliac spine** (B). The **anterior inferior iliac spine** (C) cannot readily be felt in the living person. The conjoined rami (E) have fused, and immediately superior to them is the obturator foramen (F). No evidence of the cartilage seen in the child's acetabulum remains in the adult acetabulum (D). The bony prominence of the ischium (G), called the **ischial tuberosity**, is what we actually sit on. Posterior to the ischial tuberosity, a bony projection (I) called the **ischial spine** separates two indentations in the bone. The superior and larger one (J) is called the **greater sciatic notch**; the other (H) is called the **lesser sciatic notch.** The ilium, besides having two anterior bony projections (B and C), has two posterior projections: the **posterior superior iliac spine** (L) and the **posterior inferior iliac spine** (K). These spines mark the site of the **sacro-iliac joint.** S marks the **sacrum.**

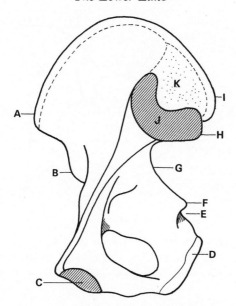

Fig. 67. Medial aspect of the hip bone

The inner side or medial surface of the hip bone (Fig. 67) has several features that cannot be seen from the lateral aspect. Orientate yourself by identifying the anterior superior iliac spine (A) and the inferior spine (B). The area C on the pubic bone is joined to the opposite pubic bone by fibrocartilage, forming the **pubic symphysis**. The ischial tuberosity (D), the lesser sciatic notch (E), the ischial spine (F), the greater sciatic notch (G), and the posterior inferior (H) and posterior superior (I) iliac spines have all been identified previously. The areas J and K articulate with the sacrum to form the sacro-iliac joint. The sacro-iliac joint is partly a synovial and partly a fibrous joint. The synovial part (J) is smooth, but the roughened area (K) is joined to the sacrum by tough fibrous connective tissue.

In Fig. 68 the right and left hip bones, seen in an anterior view, are in their normal position. The two pubic bones are held together by fibro-cartilage (A) called the pubic symphysis. The two iliac bones unite with the sacrum (I) at the right and left sacro-iliac joints (J). The several bony segments (H) at the inferior end of the sacrum form the **coccyx**. The sacrum and coccyx, together with the two hip bones, comprise the **bony pelvis**. The pelvis may be thought of as a funnel in which the lateral and posterior walls are high but the anterior wall is very short.

Other structures visible in Fig. 68 are the conjoined rami (B), the superior rami of the pubis (C), the anterior superior iliac spines (F),

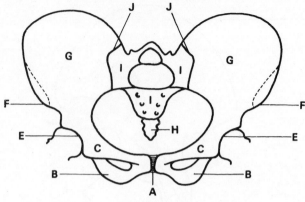

Fig. 68. Pelvis

the acetabula (E) and the iliac fossae (G). All these structures are paired: one on the right and its mirror image on the left. (Remember that *fossae* is the plural of *fossa* and *acetabula* is the plural of *acetabulum.*)

The Femur

Before looking at the hip joint, we should examine some prominent features of the femur. The right femur is shown in an anterior view in Fig. 69*a* and in a posterior view in Fig. 69*b*. The **head** of the femur (A)

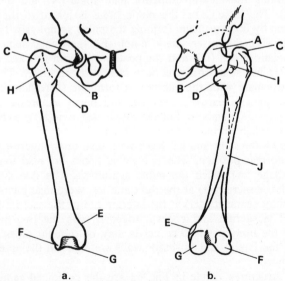

a.   b.

Fig. 69. Femur

fits into the acetabulum of the hip bone to form the **hip joint**. The **neck** of the femur (B) makes an angle of about 125° with a line drawn through the middle of the shaft of the femur. The enlargement (C) at the upper end of the shaft is called the **greater trochanter**; the smaller bony prominence (D) is called the **lesser trochanter**. A slightly raised bony ridge on the anterior aspect (H) runs from the greater to the lesser trochanter. This is called the **intertrochanteric line**; it indicates the attachment of the capsule of the hip joint.

On the posterior aspect of the femur, a well-defined ridge (I) joins the trochanters and is called the **intertrochanteric crest**. Another rough and well-defined ridge (J) runs down the posterior aspect at the shaft of the femur. It is called the **linea aspera** and indicates the attachments of several muscles. At the distal end of the shaft on the medial side, the **adductor tubercle** (E) receives the insertion of the powerful adductor magnus muscle. The medial (G) and lateral (F) **condyles** of the femur are the articular surfaces that form part of the knee joint.

THE HIP JOINT

A side view of the hip bone, with the head of the femur removed from the acetabulum, is shown in Fig. 70. For purposes of orientation the pubic symphysis is indicated at H. The acetabulum (A) is a rather deep bony cup with a gap in the inferior part of the bony rim. This gap is filled by a band of fibrous connective tissue (B) called the **transverse acetabular ligament**. From the inner edge of this ligament there arises a short round ligament (C) called the **ligamentum teres** (*teres* means 'round'). The ligament, although cut in this diagram, attaches to the head of the femur (F); the cut end of the ligament is shown at E.

A small artery (D) arises from the obturator artery and runs in the ligamentum teres to supply a part of the bony head of the femur. Other

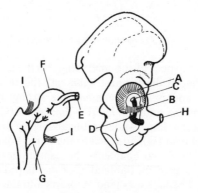

Fig. 70. Hip joint

arteries (G), arising from the femoral circumflex arteries, run along the neck of the femur and enter the bone to supply both the femoral neck and femoral head. Later, we will see how fractures of the femoral neck or dislocations of the hip joint may cut off this blood supply, resulting in necrosis or death of the bony head of the femur. Part of the capsule (I) of the joint is shown where it is attached to the femur.

Since the hip joint must bear the weight of the body during walking or standing, we need to consider how the structure of this joint provides strength and stability. Contrast the hip joint with the gleno-humeral joint. Pliable structures surrounding the gleno-humeral joint ensure its mobility. However, it has to pay the price for its lack of stability: the relative ease of shoulder dislocation. The bony acetabulum of the hip joint receives the major part of the head of the femur, thereby conferring a great deal of stability at the expense of mobility. In addition, the capsule of the hip joint is thickened in various parts to form strong ligaments.

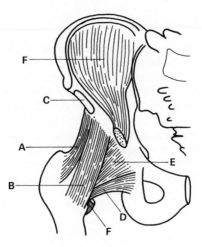

Fig. 71. Anterior aspect of hip joint

On the anterior aspect of the joint (Fig. 71) the capsule is thickened to form the Y-shaped **ilio-femoral ligament**; the two arms of the 'Y' are shown at A and B. This ligament, the strongest ligament of the body, arises from the anterior inferior iliac spine (C) and attaches to the intertrochanteric line. Another thickening of the capsule is the **pubo-femoral ligament** (D), attached to the pubic bone as well as to the femur. Between these two ligaments the capsule is relatively thin (E), but this area is covered by the tendon of the **iliopsoas muscle** (F), cut away in Fig. 71. The iliopsoas tendon inserts into the lesser trochanter. You can appreciate that the deep bony cup plus the strong ligaments

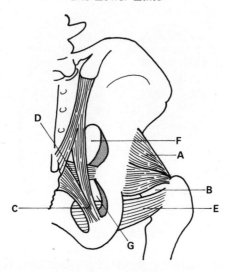

Fig. 72. Posterior aspect of hip joint

and the iliopsoas tendon confer great strength and stability to the joint. On the posterior aspect of the joint (Fig. 72) the capsule is thickened to form the **ischio-femoral ligament** (A), but it leaves uncovered the lateral part of the femoral neck (B).

Gluteal Region

Several muscles acting at the hip joint are found in the buttock, or gluteal region as it is sometimes called. The bony skeleton of the gluteal region is shown in Fig. 72. Two ligaments complete the framework: the **sacro-tuberous ligament** (C) running from the dorsal part of the sacrum to the ischial tuberosity, and the **sacro-spinous ligament** (D) running from the lower end of the sacrum and coccyx to the ischial spine. These two ligaments convert the sciatic notches into **foramina**, or doorways, through which muscles, nerves and vessels pass. Note the **greater sciatic foramen** (F) and the **lesser sciatic foramen** (G). The other structure shown in Fig. 72 is the obturator externus muscle (E), which arises from the obturator foramen and passes just below the hip joint to be inserted into the back of the femur. It laterally rotates the thigh.

Muscles of the Gluteal Region

The key muscle in the gluteal region (Fig. 73) is the **piriformis** (A), which arises from the anterior surface of the sacrum, passes through the greater sciatic foramen and is inserted into the greater trochanter. Appearing at its superior border are the **superior gluteal vessels** (B)

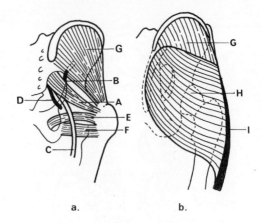

Fig. 73. Gluteal region

and nerve. At its inferior border the large **sciatic nerve** (C) emerges and passes inferior to the piriformis into the thigh. Medial to the sciatic nerve are the **inferior gluteal vessels** (D) and nerve. The **obturator internus muscle** (E) arises from the pelvic surface of the pubic bone, emerges through the lesser sciatic foramen and inserts into the greater trochanter. The short **quadratus femoris muscle** (F) arises from the ischial tuberosity and inserts into the back of the femur. All of these muscles produce lateral rotation of the femur.

Two muscles arise from the lateral surface of the hip bone and insert into the greater trochanter. The deeper **gluteus minimus** is hidden by the more superficial **gluteus medius** (G). These two muscles produce abduction of the thigh. The most massive muscle of the buttock is the **gluteus maximus** (labelled H in Fig. 73*b*), which arises from the ilium and sacrum. Most of its fibres are inserted into the **ilio-tibial tract** (I), which is the thickened lateral part of the deep fascial covering of the thigh muscles. A few of the deeper muscle fibres are inserted into the femur. The gluteus maximus muscle extends the thigh and also rotates the thigh laterally. A part of the gluteus medius muscle (G) can also be seen in Fig. 73*b*, before it passes deep to the gluteus maximus muscle.

Nerve Supply to Muscles

The nerve supply to all the muscles described in the gluteal region is derived from the **sacral plexus**. A person with one paralysed gluteus maximus muscle cannot walk up stairs or up an inclined plane. The gait of such a person is characteristic. When he bears his weight on the paralysed limb, his body and arms are suddenly pulled back. He then makes a hurried step with the healthy limb so that the weight can quickly be taken off the affected limb. If both glutei maximi muscles are paralysed, he cannot rise from a sitting position. He walks with his trunk thrust back to keep his centre of gravity behind the hip joint. When his centre of gravity falls in front of the hip joint, he is unable to prevent his trunk from bending forward at the hips.

An important function of the glutei medius and minimus muscles is shown in Fig. 74. The anterior superior iliac spines of a person standing on level ground (Fig. 74a) would lie in the same transverse plane (A). Similarly, both knee joints would lie in the horizontal plane (B). Just before the right limb takes a step the foot must clear the ground (Fig. 74b). The left glutei medius and minimus muscles (G) contract to tilt the pelvis to the left. The tilting of the pelvis enables the right foot to

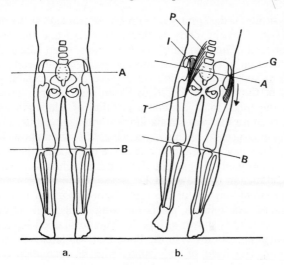

Fig. 74. Action of gluteal muscles

clear the ground so that a forward step can be taken. Neither the anterior superior iliac spines (A) nor the knee joints (B) now lie in a horizontal plane.

Perhaps the most powerful muscle that flexes the thigh is the iliopsoas (Fig. 74b). It is made up of the **psoas major muscle** (P), which arises from the lumbar vertebrae, and the **iliacus muscle** (I), which arises from the iliac fossa. The tendons of the two muscles fuse to form one tendon that is inserted on the lesser trochanter (T).

Injuries to the Hip Joint

The most common type of dislocation of the hip joint is a posterior dislocation of the head of the femur. It is usually caused by a violent force, and is often the result of a road accident. Typically, the

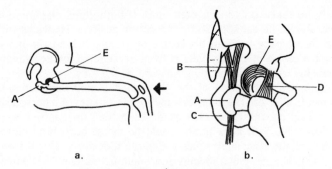

a.   b.

Fig. 75. Posterior dislocation of the hip

victim is sitting in the front seat with knees crossed; at the moment of impact (Fig. 75a) the body is thrown forward and the knee strikes the dashboard. The resultant force is transmitted along the femur in the direction of the arrow, forcing the femoral head out of the acetabulum and often breaking off a piece of the acetabular rim. As the head of the femur is forced to lie on the posterior aspect of the ischium, the thigh must assume a position of adduction and medial rotation, because the very strong ilio-femoral ligament usually remains intact. An enlargement of this region (Fig. 75b) shows how the head of the femur (A) may crush the sciatic nerve (B) against the ischium (C). The strong ilio-femoral ligament (D) is still attached to the femur. The empty acetabulum is shown at E.

Congenital dislocation of the hip is more common in female than male infants. An infant's pelvis showing congenital dislocation of the right hip joint is shown in Fig. 76. The normal left hip joint shows the growing depression (A) in the bone that will later form the cuplike acetabulum. The head of the femur, with its secondary ossification centre (B), fits into the depression. On the right side there is a very

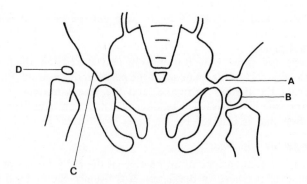

Fig. 76. Congenital dislocation of the hip

shallow depression (C) for the head of the femur. In addition, the ossification centre (D) is deformed. As the infant attempts to stand or walk, the hip joint becomes dislocated. The deformed femoral head, in attempting to support the weight of the body, slides past the abnormally shallow acetabulum. As a result, the right femur is displaced both upwards and laterally.

Fracture of the femoral neck is a frequent type of hip injury in the elderly. A fall or even a stubbing of the toe may fracture the neck of the femur. Bones of elderly people often lose calcium salts, with the result that they are unable to withstand those stresses and strains that more youthful bones can endure. In earlier days the victim was usually confined to bed. As a result of inactivity, death often occurred from lung infections or blood clots in the venous circulation. Today, early activity following the fracture is made possible by mechanically fixing the broken bone. The type of mechanical fixation most frequently used

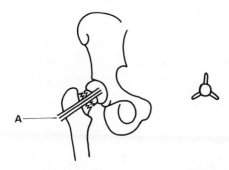

Fig. 77. Fracture of the neck of the femur

is the **Smith-Petersen nail** (Fig. 77). Through an incision in the lateral side of the thigh, the nail (A) is driven inside the neck of the femur across the fracture line so that both fragments are fixed. The thigh can then be moved without causing movement between the broken fragments of the femoral neck. The fracture will not heal if there is movement of the broken ends of the bone. The patient may even be permitted to sit in a chair a short time after the operation. An end view of the three-flanged Smith-Petersen nail is shown in Fig. 77.

Muscles of the Thigh

The general plan of the muscles of the thigh is illustrated in Fig. 78. This is a cross section through the thigh showing the cut surface of the femur (A). Grouped around the femur are three separate compartments, each containing muscles that are surrounded by a fascial stocking. The **extensor compartment** (B) is on the anterior aspect of the thigh. This compartment contains muscles whose function is to extend

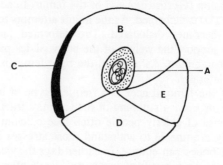

Fig. 78. Muscular compartments of thigh

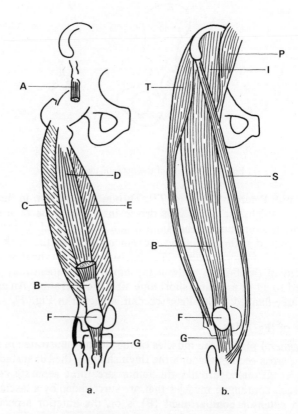

Fig. 79. Muscles of thigh (anterior)

the leg. The fascial stocking covering the lateral aspect of the extensor compartment is thickened and is called the **ilio-tibial tract** (C). It is so named because it extends from the ilium, the bone of the pelvis, to the tibia. On the posterior aspect of the thigh is the **flexor compartment** (D) which contains the muscles that flex the leg. On the medial side of the thigh is the **adductor compartment** (E) which contains the muscles that adduct the thigh, not the leg.

The four muscles of the extensor compartment are shown in Fig. 79*a*. The longest one, the **rectus femoris**, arises from the anterior inferior iliac spine (A) and is cut to show the underlying muscle. The other end of the muscle (B) inserts into the patella (F). The remaining three muscles are grouped around the femur. The **vastus lateralis** (C), the **vastus medialis** (D) and the **vastus intermedius** (E) all arise from the femur and insert into the patella. From the patella a ligament called the **patellar ligament** (G) inserts into the tibia. When these muscles contract, they extend the leg.

Two muscles not considered part of the extensor compartment are shown in Fig. 79*b*. The **sartorius muscle** (S) arises from the anterior superior iliac spine and is inserted in the tibia. The **tensor fascia lata** (T) arises from the ilium and inserts into the lateral fascia of the thigh. As mentioned above, this lateral fascia is thickened and called the ilio-tibial tract. Both the tensor fascia lata and the sartorius muscles flex the thigh.

Two other flexors of the hip joint are shown in Fig. 79*b*. They are the **psoas** (P) arising from the lumbar vertebrae and inserted into the lesser trochanter of the femur, and the **iliacus** (I) arising from the iliac fossa of the hip bone and inserted into the lateral side of the psoas tendon.

The adductor muscles of the thigh are shown in Fig. 80. The anterior aspect of the right thigh is shown in Fig. 80*a* and the posterior aspect in Fig. 80*b*.

The **pectineus** (A) and **adductor longus** (C) lie in the same plane. Just deep to the adductor longus is the **adductor brevis muscle** (B). The deepest muscle is the **adductor magnus** (D). Its insertion is shown more clearly in the posterior view. The long medial border is inserted into the linea aspera on the posterior surface of the femur up to the level of the lesser trochanter and down to the adductor tubercle. Where the inferior part of the muscle meets the femur, there is a gap (E) in the muscle. This opening allows the passage of the large femoral artery from the anterior aspect of the thigh to the posterior aspect of the knee joint. The **gracilis** (F), a slender muscle arising with the other adductors, is inserted on to the tibia. All these muscles are used to adduct the thigh.

Fig. 80*b* shows that the adductor magnus consists of a transverse part (D) and a longitudinal part (G), which arises from the ischial tuberosity (H) and is inserted into the adductor tubercle. Because of the direction of these fibres, their action is to extend the thigh. All the

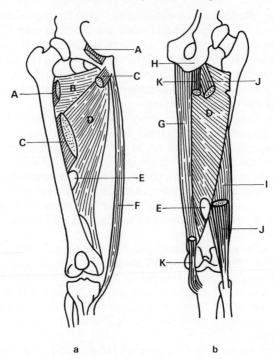

a   b

Fig. 80. Muscles of thigh (posterior)

muscles which arise from the ischial tuberosity are called the **hamstrings**. They consist of the longitudinal fibres of the adductor magnus (G), the long head of the **biceps femoris** (J) and two other muscles shown together in Fig. 80*b* as K. These two muscles consist of the broader **semimembranosus** which is inserted on to the back of the tibia, and the thinner **semitendinosus** which is inserted on to the antero-medial surface of the tibia in common with the sartorius and gracilis muscles. The biceps femoris consists of two parts, the **long head** (J) which arises from the ischial tuberosity, and the **short head** (I) which arises from the lower part of the linea aspera of the femur. The two heads join to form a common tendon which is inserted into the head of the fibula.

The long head of the biceps, the semimembranosus and the semitendinosus produce movement at two joints: they extend the thigh at the hip joint and flex the leg at the knee joint. They hinder flexion at the hip joint, when the knees are extended, by their passive resistance. Lie flat on your back and try raising your legs, keeping the knees straight. The limit of movement is usually reached when the lower limbs are vertical, but there are great individual differences due to the effects of training or lack of exercise.

THE KNEE JOINT

The knee joint is the largest joint in the body because it is formed by two of the largest bones, the femur and the tibia. Although the knee is subject to considerable strain, it does not have the advantage of a deep bony socket, as is found in the hip joint. It depends primarily on muscles and ligaments for its stability.

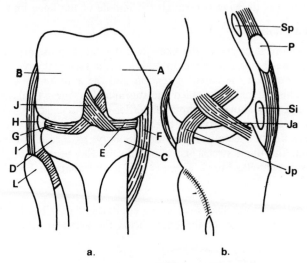

a.   b.

Fig. 81. Knee joint (anterior and lateral)

The general features of the joint are illustrated in Fig. 81, with the anterior view shown in Fig. 81*a* and the lateral view in Fig. 81*b*. The rounded medial (A) and lateral (B) condyles of the femur are separated from the medial (C) and lateral (D) condyles of the tibia by two wedge-shaped discs of cartilage called **menisci** (singular, *meniscus*). The **medial disc** (E) is firmly bound to the fibrous capsule (F) of the joint. The **lateral disc** (G) is more mobile and is not so closely bound to the capsule (H). The **lateral ligament** (I) of the knee joint is a separate cord, not a thickening of the capsule, and is attached to the head of the fibula (L).

On the upper surface of the tibia, between the condyles, there is a central region called the **intercondylar eminence.** Two cordlike ligaments arise from this region and cross one another as they run to the intercondylar notch of the femur. The two ligaments cross like the arms of an X and are called the **cruciate ligaments** (J). They are the internal ligaments of the knee joint, whereas the lateral ligament (I) is an external ligament.

The lateral view of the joint (Fig. 81*b*) shows the **patella** (P) which receives the tendons from the four extensor muscles and is anchored to the tibia by a continuation of this tendon. The anterior cruciate ligament (Ja) arises anterior to the central region of the tibia and passes upwards and backwards, to be attached to the lateral condyle of the femur. The posterior cruciate ligament (Jp) arises posterior to the central region and passes upwards and forwards to the medial condyle. Both ligaments are inserted into the sides of the intercondylar notch of the femur (the drawing is through the middle of the knee joint).

Synovial Cavity

The synovial cavity of the knee joint is related to the articular surfaces of the femur and tibia. In addition, the synovial cavity extends upwards between the extensor tendon and the anterior surface of the femur. This extension is called the **suprapatellar bursa** (Sp), and may extend 2 inches above the superior border of the patella. Stab wounds in this area may introduce infection into the synovial cavity. If excessive fluid collects in the synovial cavity, as a result of injury or infection, the thigh swells just above the patella because of the fluid in the suprapatellar bursa. There is also an **infrapatellar bursa** (Si).

Menisci

The cartilaginous discs rest on the condyles of the tibia. In Fig. 82*a* the transverse section of the knee shows the articular surface of the

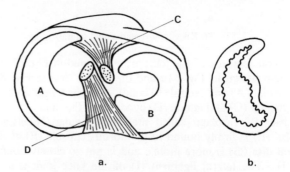

Fig. 82. Menisci and cruciate ligaments

tibia. The medial disc (A) is larger and less curved than the lateral disc (B). They both resemble a half-moon in outline and are sometimes called the **semilunar cartilages**. The discs are attached to the tibia by their pointed ends, called horns. The anterior horns are separated by the origin of the anterior cruciate ligament (C) and the posterior horns by the origin of the posterior cruciate ligament (D). The condyles of the femur are in contact with the discs, which act as shock-absorbers during walking and jumping. There is movement between the menisci and the articular surfaces of both the femur and the tibia during flexion and extension of the knee joint.

Athletes frequently injure a meniscus, the medial meniscus being most often affected. A typical history of such an injury is that of a footballer who twists his flexed knee while running and feels a sudden pain accompanied by 'locking' of the knee joint. The joint cannot be extended and cannot bear weight. The medial meniscus is drawn to the centre of the joint and is crushed by the medial femoral condyle during attempted extension of the twisted joint. Fig. 82*b* shows a typical tear of the medial meniscus. Since a tear in fibrocartilage does not heal, it may be necessary to remove the damaged meniscus.

Popliteus Muscle

In Fig. 83 the medial meniscus (A) is attached to the capsule (B) as seen in this posterior view of the right knee joint. The lateral menis-

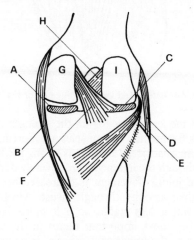

Fig. 83. Knee joint (posterior)

cus (C) is separated from the capsule (D) by the tendon of the **popliteus muscle** (E). This muscle medially rotates the tibia, if the foot is off the ground, or it laterally rotates the femur when the leg is bearing the body's weight. The posterior cruciate ligament (F) is shown attached to the medial femoral condyle (G), and the anterior cruciate (H) is attached to the lateral femoral condyle (I).

Extensor Muscles

In Fig. 84 the four extensor muscles are seen at their insertion. The vastus lateralis (A), vastus medialis (B), vastus intermedius (C) and the cut portion of rectus femoris (D) constitute the four heads of the **quadriceps.** The four heads form one tendon (F) which is inserted into the anterior aspect of the tibia. The patella (E) is inserted into this tendon. Arising from the vastus lateralis and the vastus medialis are sheets of strong connective tissue (G) which form part of the capsule of the knee joint. These sheets of tissue are called the **patellar retinacula.**

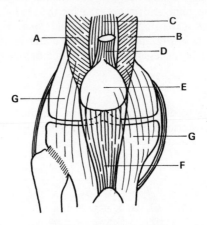

Fig. 84. Patella and insertion of extensor tendon

The stability of the knee depends on the strength of its surrounding muscles and of its ligaments. The muscles are more important than the ligaments, for if the quadriceps femoris is strong the knee can continue to function even though the ligaments are severely damaged. Following injury to the knee joint, movement is decreased because of pain; the quadriceps may then atrophy through disuse. Atrophy of the quadriceps means the joint is less stable, so it is more likely to be injured again. Probably the most important factor in the treatment of knee injuries is exercise of the quadriceps muscle. How would you exercise this muscle without bending the knee?

THE LEG

The muscles of the posterior aspect of the leg are shown in Fig. 85. Fig. 85*a* shows the deeper muscles. The **tibialis posterior** (A) is over-lapped on the medial side by the **flexor digitorum longus** (B) and on the lateral side of the **flexor hallucis longus** (C). The tendons cross the ankle joint and reach the foot (see page 89). The tendon of the

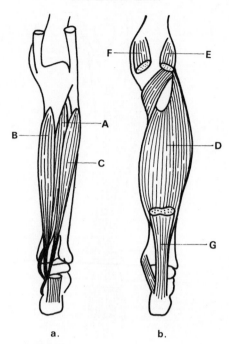

a. b.

Fig. 85. Muscles of calf (superficial and deep)

tibialis posterior lies in a groove of the medial malleolus, and the other two pass under a projection of the calcaneum.

Fig. 85*b* shows the two superficial muscles which form the calf of the leg. The larger and deeper muscle is the massive **soleus** (D), which arises from both tibia and fibula. The **lateral head** of the **gastrocnemius** (E) arises just above the lateral condyle of the femur, and the **medial head** (F) arises just above the medial condyle. They unite to form the gastrocnemius muscle. The gastrocnemius and the soleus unite to form a common tendon (G) known as the **Achilles tendon**, which is inserted on the **calcaneum** (heel bone). The action of the calf muscles is chiefly at the ankle joint in plantar flexion of the foot (see page 61).

The muscles of the anterior aspect of the leg are shown in Fig. 86. The **tibialis anterior** (A) and the **extensor digitorum** (B) **muscles** partially cover the **extensor hallucis longus** (C). The tendon of the tibialis anterior is inserted on the foot (D), while the other tendons are inserted on the toes. E represents two muscles on the lateral aspect of the leg: the **peroneus longus** and the **peroneus brevis**. These two muscles arise from the lateral surface of the fibula and their tendons pass behind the lateral malleolus (F) to reach the foot. (The insertion of the tendons on to the foot is discussed on page 87.)

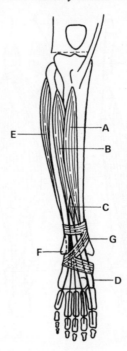

Fig. 86. Muscles of leg (anterior)

The muscles of the leg are surrounded by connective tissue called fascia. It is continuous above with the fascia lata of the thigh and blends with the periosteum on the medial surface of the tibia, which is just under the skin. It is attached to the underlying muscles and acts as a kind of elastic stocking for the structures of the leg. In the region of the ankle it is thickened to form bands, called **retinacula**, which serve to maintain the tendons in position as they cross the joint. The extensor retinacula (G) are shown at the front of the ankle; the superior is a broad sheet which crosses the extensor tendons and is attached to the lower parts of the tibia and fibula. The lower retinaculum is shaped like a 'Y' and curves round the instep.

NERVES AND VESSELS OF THE LOWER LIMB

The major artery of the lower limb is the **femoral artery**, shown in Fig. 87. At the root of the limb, the **external iliac artery** (A) becomes the femoral artery (B). It continues distally to pass through the opening (D) in the adductor magnus muscle to reach the back of the knee joint, where it is called the **popliteal artery**. It gives off three relatively small branches in the groin (not shown in Fig. 87). About two inches from its origin it gives off a large branch called the **profunda femoris** (C). This gives off medial and lateral **circumflex** branches which supply the hip joint, and the lateral circumflex sends down a branch to join the collateral circulation of the knee joint. The profunda femoris passes distally to supply the deep muscles of the thigh.

Emerging from the obturator foramen of the pelvis is the **obturator artery**, which sends one branch (E) to the hip joint and another branch

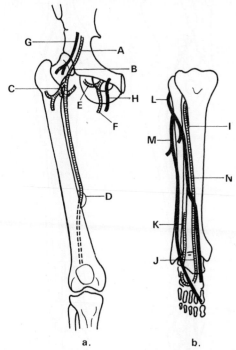

a.   b.

Fig. 87. Arteries and nerves of lower limb (anterior)

(F) to the adductor muscles of the thigh. The nerve (H) emerging from the obturator foramen is the **obturator nerve,** the chief motor nerve supply to the obturator muscles. The nerve lying lateral to the femoral artery is the **femoral nerve** (G). It supplies the quadriceps muscle and the sartorius muscle. Other branches supply the skin on the anterior and medial surfaces of the thigh.

Fig. 87*b* shows the arteries and nerves on the anterior aspect of the leg and foot. The **anterior tibial artery** (I) arises from the popliteal artery and passes forward between the tibia and fibula to descend in the anterior compartment of the leg. When it crosses the ankle joint, it becomes the **dorsalis pedis artery** (J) which runs forward on the dorsal surface of the foot and then plunges between the first and second metatarsals. The distal end of the **peroneal artery** (K) emerges between the leg bones and runs on the dorsal surface of the foot to form an arch with the dorsalis pedis artery.

The **common peroneal nerve** (L) appears on the lateral aspect of the fibula, where it soon divides into a superficial peroneal branch (M)

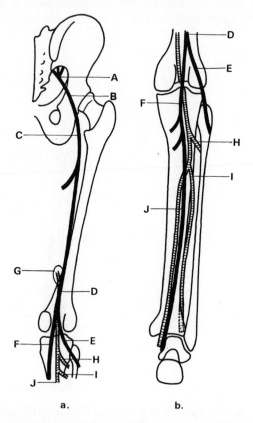

Fig. 88. Arteries and nerves of lower limb (posterior)

and a deep peroneal branch (N). The superficial nerve supplies the peroneal muscles and continues distally in the leg and foot, where it supplies skin. The deep peroneal branch supplies the tibialis anterior, extensor hallucis longus and extensor digitorum longus muscles.

In the extensor compartment of the thigh, Fig. 88a, the large **sciatic nerve** arises in the pelvis and passes through the greater sciatic foramen (A) to lie on the ischium (B), where it is liable to injury in posterior dislocations of the hip joint. It passes midway between the ischial tuberosity and the lesser trochanter (C). It gives muscular branches to all the extensor muscles, including the hamstring part of the adductor magnus muscle.

Above the knee joint the sciatic nerve (D) divides into a lateral or **common peroneal branch** (E) and a medial or **tibial branch** (F). The common peroneal nerve passes round the posterior surface of the head of the fibula, where you can feel it by rolling it under your finger against the bone. It continues round the fibula to reach the anterior surface of the leg. The tibial nerve (F) supplies all the flexor muscles in the posterior compartment of the leg.

As soon as the femoral artery emerges through the opening in the adductor magnus muscle (G) it is called the popliteal artery. After giving off the anterior tibial artery (H), which passes between the tibia and fibula to the front of the leg, the popliteal becomes the **posterior tibial artery**. This gives rise to the peroneal artery (I), which supplies muscles in the back of the leg. The posterior tibial artery (J) passes distally to reach the foot.

In Fig. 88*b* the vessels and nerves at the back of the leg are continued. The peroneal artery (I) is shown passing distally just above the ankle joint where it passes between the leg bones to the anterior part of the ankle. The posterior tibial artery (J) crosses the posterior aspect of the ankle joint to reach the foot.

The sciatic nerve (D) divides into the common peroneal nerve (E) and the tibial nerve (F). The tibial nerve gives branches to the gastrocnemius and soleus muscles and also to the tibialis posterior, flexor digitorum longus and flexor hallucis longus muscles. It continues distally to accompany the posterior tibial artery to the foot.

The veins of the lower limb are divided into deep and superficial groups. The deep veins accompany the corresponding arteries and lie deep to the fascia which forms an 'elastic stocking' to the limb. The superficial veins lie just under the skin and convey blood to the deep veins. They are called the long and short saphenous veins (Fig. 89). The **long saphenous vein** (labelled B in Fig. 89*a*) drains the dorsum of the foot and passes upwards just in front of the **medial malleolus** (A) to the groin, where it pierces the fascia an inch below the **inguinal ligament** (C) to enter the **femoral vein** (D). In addition to this main junction there are numerous communicating channels between the deep femoral vein and the superficial saphenous vein. Both veins have valves which permit the blood to flow one way only—towards the heart.

The **short saphenous vein** (Fig. 89*b*) commences behind the lateral malleolus of the ankle and passes up over the back of the calf (F). It perforates the deep fascia behind the knee joint and joins the **popliteal vein** (G), which becomes the femoral vein (D) higher up the thigh. The femoral vein becomes the external iliac vein (E) in the pelvis.

The long saphenous vein frequently becomes **varicose**, i.e. dilated

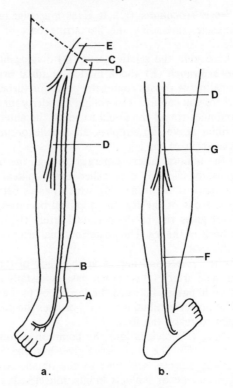

Fig. 89. Veins of lower limb

and twisted. When a person with varicose veins stands upright for a short time, the dilated veins fill with blood and are very prominent. The valves become incompetent and so blood can flow back down the leg and also pass from the deep veins back into the superficial veins. Pregnancy and standing for long periods cause increased venous pressure which aggravates varicose veins. Elevation of the lower limbs and elastic stockings can ease the discomfort.

THE FOOT

The bones of the foot are shown in Fig. 90 from the medial side of the foot; in Fig. 91 from the lateral side of the foot; and in Fig. 92 from the plantar aspect or sole. The bones of the foot are called **tarsal** and **metatarsal** and correspond to the carpal and metacarpal bones of the hand.

The **talus** (A) articulates with the distal ends of the tibia and fibula to form the ankle joint. The largest tarsal bone is the **calcaneum** (B).

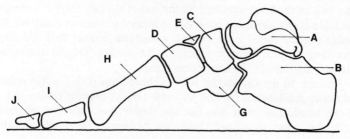

Fig. 90. Bones of foot (medial)

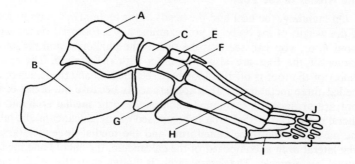

Fig. 91. Bones of foot (lateral)

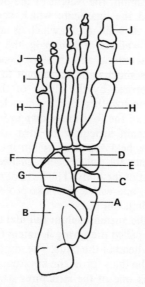

Fig. 92. Bones of foot (plantar)

The heel of the foot is formed by the end of the calcaneum. Anterior to the talus is the **navicular bone** (C). Anterior to the navicular are three wedge-shaped tarsal bones called **cuneiform bones**; they are named according to their relative positions, the medial (D), the intermediate (E) and the lateral (F) cuneiform bones.

Anterior to the calcaneum is the **cuboid bone** (G). The five **metatarsal** bones are labelled H. The phalanges of the toes are similar to those of the hand. The great toe, like the thumb, has only two phalanges, while the other toes have three: the proximal (I), middle and distal (J) phalanges.

The Arches of the Foot

On standing, the heel and the heads of the metatarsals support most of the weight of the body. If you examine a wet footprint on the bathroom floor, you can see that the tips of the phalanges and the lateral border of the foot are also in contact with the ground. The medial aspect of the foot is off the floor between the heel and the heads of the medial three metatarsals. This appearance is because the bones of the foot are arranged in two longitudinal arches: the **medial arch** and the **lateral arch**. The medial arch is composed of the calcaneum, the talus, the navicular, the three cuneiforms and the medial three metatarsals. The lateral arch is composed of the calcaneum, the cuboid and the two lateral metatarsals. The lateral arch is flatter than the medial arch, which explains the shape of the wet footprint.

In the standing position, the weight of the body locks the bones of the foot together and the ligaments which connect them are under tension. The foot is a solid arch, which is an efficient method of supporting weight. In walking and running, the weight is released from the arches, the bones unlock and the foot becomes a flexible, active spring, which absorbs shocks and can adapt to uneven ground.

The arches are formed by the shape of the bones, by the ligaments which hold them together, and by muscle action. The ligaments on the plantar aspect of the foot are shown in Fig. 93. The body of the talus and the calcaneum articulate to form the **talo-calcaneal joint.** These two bones are held together by an interosseus ligament (not shown) which prevents the talus being pushed forward on the calcaneum by the weight of the body. The head of the talus (A) articulates with the navicular (B) and with a shelf-like projection of the calcaneum called the **sustentaculum tali** (C).

The gap between the sustentaculum tali and the navicular is bridged by a strong ligament called the **spring ligament** (D), which supports the inferior aspect of the head of the talus. The importance of this ligament is apparent because the talus forms the keystone of the medial arch and the spring ligament is one of the structures which help to prevent the talus being forced towards the ground by the weight of the body. The

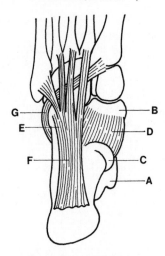

Fig. 93. Plantar ligaments

calcaneum also forms a joint with the cuboid bone (E). Since the lateral arch of the foot may be depressed at this point, a strong ligament is found on its plantar aspect. This is the **long plantar ligament** (F).

The joints which the calcaneum makes with the three bones mentioned above (head of the talus, the navicular and the cuboid) are together referred to as the **transverse tarsal joint**, since the joint lines lie approximately in the same transverse plane. During inversion and eversion of the foot (see Fig. 65) movement takes place at this joint as well as at the talo-calcaneal joint.

Foot disorders are very common and one frequently hears the term 'flat feet', which means depression of the medial longitudinal arches. It is usually associated with some degree of eversion. All infants have flat feet for a year or two after they begin to stand. When the deformity does not correct itself, the tarsal bones grow into shapes which form a straight line when articulated instead of an arch. The ligaments may then become strained by the ordinary amount of standing which is part of everyday life. In the normal foot, the ligaments are protected from strain by the action of the muscles which maintain the arches of the foot. In the adult flat foot these muscles are not maintaining the arches and so more stress falls on the ligaments. This is the cause of the pain which is felt on standing too long on flat feet.

Tendons and Muscles of the Foot

The deepest layer of tendons and muscles of the foot is shown in Fig. 94. The **peroneus longus tendon** (A) arises from the lateral side of the leg and bends round the lateral border of the foot. It is partially

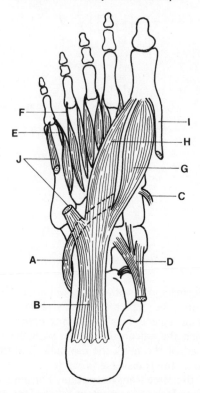

Fig. 94. Muscles of foot (deep layer)

covered by the long plantar ligament (B) and is inserted on the base of the first metatarsal and adjacent medial cuneiform. This tendon is shown more clearly as G in Fig. 93. It everts and plantar-flexes the foot. The **tibialis anterior tendon** (C) is inserted on the opposite side of the same two bones. This tendon passes anterior to the ankle joint so its contraction inverts and dorsiflexes the foot. The **tibialis posterior tendon** (D) divides into slips that are inserted on to several of the tarsal bones. The tibialis posterior muscle arises from the posterior aspect of the leg and uses the medial malleolus as a pulley to plantar-flex and evert the foot.

The interossei muscles in the foot have the same general plan as in the hand. Since the toes have little independent movement, they act together to flex the metatarso-phalangeal joints. They probably prevent displacement of the metatarsal bones during weight-bearing, and they are concerned with the mechanism of the arches of the foot. In Fig. 94 one of the plantar interossei is labelled E, and one of the dorsal interossei is labelled F.

A group of three muscles lies just superficial to these structures. They arise from the tarsal bones and the long plantar ligament which covers these bones. The **flexor hallucis brevis** (G) is inserted on to the base of the proximal phalanx of the great toe, along with the tendon of the **adductor hallucis muscle** (H). The tendon of the **abductor hallucis muscle** is also shown (I), but the muscle itself is not shown in full. The **flexor digiti minimi** (J) is also shown. The names of these muscles indicate their function.

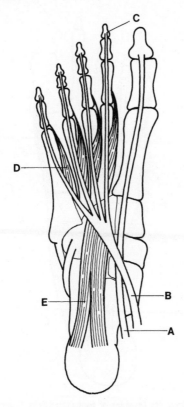

Fig. 95. Muscles of foot (intermediate layer)

Lying immediately superficial to the preceding structures are the long flexor tendons of the foot. These are shown in Fig. 95. The **flexor hallucis longus tendon** (A) passes superficially to the sustentaculum tali and the spring ligament, to be inserted on to the base of the distal phalanx of the great toe. It reinforces the spring ligament and thus aids in maintaining the medial arch of the foot. The **flexor digitorum longus tendon** (B) takes a more oblique course, and a slip from the tendon is inserted on the distal phalanx of each toe (C). As in the hand, a small muscle arises from each slip (D). These are the four lumbrical muscles, which are inserted on the extensor tendons of the lateral four toes. The **accessory flexor** (E) is inserted into the long flexor tendon, correcting for the oblique pull on the toes.

The most superficial group of muscles of the foot is shown in Fig. 96. This group is made up of the **abductor digiti minimi** (A) on the

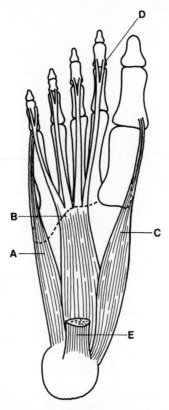

Fig. 96. Muscles of foot (superficial layer)

lateral side of the foot, the **flexor digitorum brevis** (B) and the **abductor hallucis** (C) on the medial side of the foot. They arise from the calcaneum and related fascia and are inserted as shown. The flexor digitorum brevis gives rise to four tendons, each of which splits just before inserting on to the base of the middle phalanx (D). Superficial to the flexor digitorum brevis is a thickened layer of fascia (E) called the **plantar aponeurosis,** which runs towards the toes to become attached to the fibrous sheaths surrounding the tendons in the toes. It is similar to the palmar aponeurosis in the hand. The plantar aponeurosis is cut in Fig. 96 and its thickness is immediately apparent. It is a strong fibrous support for the arches of the foot.

Nerves and Arteries of the Foot

The nerves and arteries of the sole of the foot are shown in Fig. 97. The posterior tibial artery (A) divides to form the **lateral plantar artery** (B) and the **medial plantar artery** (C). The medial plantar artery continues towards the big toe. The lateral plantar artery curves medially

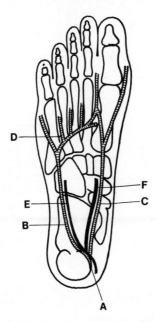

Fig. 97. Arteries and nerves of foot (plantar)

to form the **plantar arch** (D). This arch continues to the space between the first and second metatarsals, where it anastomoses with the dorsalis pedis artery from the dorsal aspect of the foot. Metatarsal branches arising from the arterial arch are seen passing to the toes.

The posterior tibial nerve, like the artery, divides into the **lateral plantar** (E) and the **medial plantar** (F) **nerves.** The medial plantar supplies the flexor digitorum brevis, the abductor hallucis and flexor hallucis muscles. The lateral plantar nerve supplies all the other muscles of the foot. The supply to the skin is shown by the line drawn down the fourth toe; the medial nerve supplies the skin medial to this line and the lateral nerve supplies the rest.

THE ANKLE JOINT

Three views of the ankle joint are shown in Fig. 98. Fig. 98*a* is a frontal section through the joint, Fig. 98*b* shows the medial aspect, and Fig. 98*c* shows the lateral aspect. The ankle joint is formed by three bones: the tibia, the fibula and the talus. The medial malleolus of the tibia (T), the distal end of the tibia and the lateral malleolus of the fibula (F) form a rigid socket for the body of the talus (Ta). Movement is that of a hinge joint, and only plantar-flexion and dorsiflexion of the

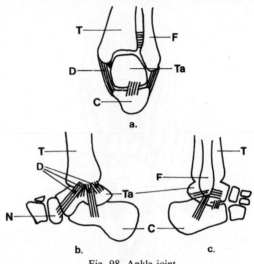

Fig. 98. Ankle joint

foot occur. While the malleoli prevent sideways movement, strong ligaments on the medial and lateral sides of the joint prevent the talus from slipping away from the leg bones.

When forcible side-to-side movements occur, the ligaments are frequently torn. The ligament on the medial side of the joint is called the **deltoid** (D). It is attached to the tip of the medial malleolus and runs to the body of the talus, the sustentaculum tali and the navicular (N). It is liable to be injured in forcible eversion of the foot, but is usually so strong that the tip of the malleolus is torn off before the ligament gives way. There are three ligaments on the lateral side of the ankle, all attached to the tip of the lateral malleolus, and then separating to become attached to the calcaneum (C) and the body of the talus (Ta). They are frequently torn when the foot is forcibly inverted.

CHAPTER SEVEN

THORAX, ABDOMEN AND PELVIS

The trunk contains the three major cavities of the body: the **thoracic cavity**, the **abdominal cavity** and the **pelvic cavity**. In a sagittal section through the trunk, the three major cavities are indicated. The thoracic cavity (A) contains the heart, the lungs, and the food and air passages. This cavity is surrounded by the ribs which form a bony cage. The ribs are attached posteriorly to the 12 **thoracic vertebrae** (T) which are

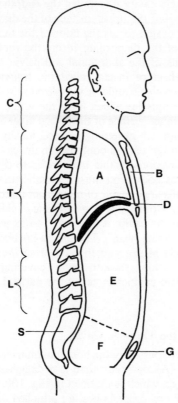

Fig. 99. Sagittal section of trunk

continuous above with the **cervical vertebrae** (C) and below with the **lumbar vertebrae** (L). The **sacral vertebrae** are fused together to form the **sacrum** (S). The ribs are attached anteriorly to the **sternum** (B). A muscular septum or wall called the **diaphragm** (D) separates the thoracic cavity from the abdominal cavity (E). Although the ribs form part of the side wall of the abdominal cavity, most of the side wall is composed of muscles attached to the lumbar vertebrae, to the lower ribs and to the bones of the pelvis. The pelvic cavity (F) is directly continuous with the abdominal cavity. A transverse plane passing from the pubic symphysis (G) to the anterior surface of the first sacral vertebra is regarded arbitrarily as the division between the abdominal and pelvic cavities.

Lining Membranes

All three cavities are lined by several layers of membrane. The outer lining membrane consists of a dense sheet of connective tissue or fascia. In the thoracic cavity it is called the **endothoracic fascia**, and it covers all the walls and the upper surface of the diaphragm. The upper surface of the diaphragm forms the floor of the thoracic cavity, while the lower surface of the diaphragm forms the roof of the abdominal cavity. All the walls of the abdominal and pelvic cavities are covered by a continuous sheet of fascia called the **transversalis fascia.** The fascia in all three cavities should be visualized as sticking closely to the deep surface of the muscles which form the walls.

Adhering to the inner surface of the fascial layer is a thin transparent membrane with a smooth glistening surface, ideally suited to lessening the friction between the organs contained in the cavities. (Although the diagrams and descriptions in this book refer to the position of structures in a motionless body, you must remember that in the *living* body almost all the organs are capable of movement relative to one another.) In the thoracic cavity this smooth membrane is called **pleura**. Because it lines the walls of the cavity it is referred to as **parietal pleura** (from the Latin *paries* meaning 'wall'). The pleura covering the diaphragm is called **diaphragmatic pleura** and that covering the side wall is called **costal pleura**. The thin, smooth layer covering the walls of the abdominal and pelvic cavities is called the **parietal peritoneum.**

THE THORAX

Compartments of the Thoracic Cavity

First, identify on your body the bony landmarks which can easily be felt. The sternum (A) or breastbone is a dagger-shaped bone at the front of the rib-cage, which is outlined in Fig. 100 by the line marked C. The clavicles (B) or collar bones are attached anteriorly to the top of the sternum. The heart (D) separates the right lung (E) and the left

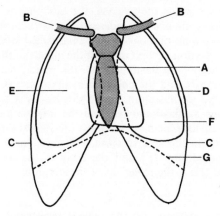

Fig. 100. Surface markings of the thorax

lung (F). The medial aspects of both lungs are behind the heart. The upper surface of the diaphragm is indicated in Fig. 100 by the dotted line G.

Fig. 101 is a transverse section of the thorax. The bony cage is formed by the sternum (A) in front, by the ribs (B) on the sides, and by

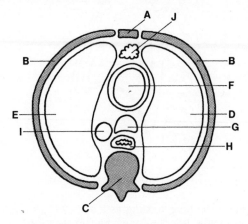

Fig. 101. Transverse section of thorax

the thoracic vertebrae (C) at the back. The right (D) and left (E) **pleural sacs** (which contain the lungs) are separated by a septum or wall called the **mediastinum**. This is formed by the heart (F), the **trachea** or air passage (G), the **oesophagus** or food passage (H) and the great artery from the heart, the **aorta** (I). The space in front of the heart is called the **anterior mediastinum** (J); it contains some fat and a few lymph nodes. The space behind the heart is called the **posterior**

mediastinum; it contains the trachea, the oesophagus and the aorta. Pleura is generally named according to the structure it covers; the pleura on the side walls of the mediastinum is called **mediastinal pleura.**

Fig. 102 is a sagittal section through the middle of the mediastinum. It shows the anterior mediastinum (A), the middle mediastinum containing the heart (B), and the posterior mediastinum (C). For descriptive purposes the mediastinum is also divided by a transverse plane into the **superior mediastinum** (D) and the **inferior mediastinum** (E). The line which divides them passes from the sternal angle to the lower border of the fourth thoracic vertebra. The sternal angle is the junction between the **manubrium** (F) and the body of the sternum (G). The **xiphoid process** (H) is the lower tip of the sternum.

The Lungs

The lungs are enclosed by the pleural sacs which form a double layer of pleura between the lung and the thoracic wall. How this comes about is shown in Fig. 103. This is a horizontal section through the

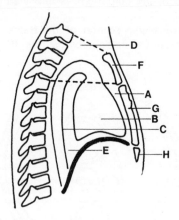

Fig. 102. Sagittal section of thorax

thorax and shows the lung starting as a bud (A) sprouting from the air passage, or **bronchus** (B). The lung bud pushes into the mediastinal pleura (C). In a later stage the lung has grown and has pushed the mediastinal pleura in front of it to form the **pleural sac** (D). The layer of pleura covering the lung is called **visceral pleura** (E). The pleura lining the thoracic wall is called the **parietal pleura** (F). The space between the visceral and parietal pleura is called the **pleural cavity** (D). Normally, the two layers are very close together and are separated only by a film of fluid. This fluid lubricates the two layers which slide on each other as the lung moves in breathing. In diseases of the lung or pleura, excess

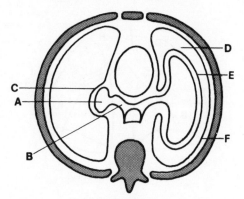

Fig. 103. Pleural sacs

fluid may collect in the pleural cavity. This compresses the lung and makes breathing more difficult. In **pleurisy**, an area of parietal pleura is infected and inflamed, which makes it rough. During breathing, the visceral pleura rubs painfully over this rough area.

Lobes of the Lung

The anterior surfaces of both lungs are shown in Fig. 104. The right lung is divided into three lobes: the **upper lobe** (A), the **middle lobe** (B)

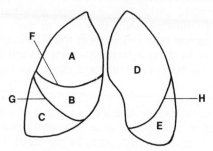

Fig. 104. Lungs (anterior)

and the **inferior lobe** (C). These lobes are separated by the two fissures: the **horizontal fissure** (F) and the **oblique fissure** (G). The left lung consists of an upper lobe (D) and a lower lobe (E), separated by an oblique fissure (H).

The medial surfaces of the lungs are shown in Fig. 105. The lobes of each lung are indicated as before. The trachea (F) divides into the right bronchus (G) and left bronchus (H). The right bronchus is more in line with the trachea and the left bronchus is pushed more laterally. Inhaled foreign objects, such as peanuts, are found in the right lung more frequently than the left. In the area where the bronchi (plural of

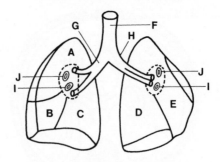

Fig. 105. Lungs (medial)

bronchus) enter the lungs, the cut ends of two other structures are seen. The **pulmonary vein** (I) carries oxygenated blood back to the heart, while the **pulmonary artery** (J) carries blood to the lungs to be oxygenated.

The oval area where all the structures either enter or leave the lung is called the **lung root**. The lung root contains some other structures which are not shown. The air passages themselves receive blood from the bronchial arteries, which are small branches of the descending aorta. The lungs have a rich network of lymphatic vessels which drain towards the lung root. Foreign particles in the inspired air pass into the lymph and are filtered out of it by the lymph glands at the root of the lung. These lymph glands become grey or even black in town-dwellers because of the carbon (soot) particles in them. Infectious bacteria are also drained away by the lymphatic system and they, too, lodge in the lymph glands at the root of the lung and may cause their enlargement.

Lung Segments

Each lung is further subdivided into a number of segments, each of which is supplied by a segmental bronchus, artery and vein. The right lung divides into ten segments: three in the upper lobe, two in the middle lobe, and five in the lower lobe. The left lung also has ten segments: five in the upper lobe and five in the lower lobe. The segments are wedge-shaped. If the lung is cut carefully along the boundaries of the segments there is very little bleeding or leakage of air from the air spaces.

Fig. 106 is a diagram of a segment. It is covered on its periphery by visceral pleura (A) and it is separated from neighbouring segments by connective tissue (B). The bronchus (C) becomes increasingly smaller as it branches, and each branch finally ends in thin-walled air-sacs called **alveoli** (D).

The pulmonary artery carrying blood to be oxygenated is closely associated with the bronchus. It sends out branches to form a network

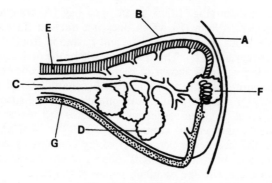

Fig. 106. Lung segment

of capillaries on the surface of each alveolus. Here, the blood in the thin-walled capillaries is very near the air in the thin-walled alveolus and the exchange of oxygen in the alveoli for carbon dioxide in the blood can occur. The oxygenated blood leaves the capillaries to enter the pulmonary vein which lies in the connective tissue separating the segments. Many diseases of the lung are at first confined to one segment, and the removal of one segment is better than the removal of a whole lobe. Equally, removal of one lobe is better than removal of a whole lung; but, if absolutely necessary, the entire lung can be removed as it is possible to live with one lung only.

The Heart

The heart and the roots of the great blood-vessels are contained in a tough fibrous bag called the **fibrous pericardium**, which is lined with a thin, glistening membrane called the **serous pericardium**. Serous pericardium is similar to pleura in both structure and function. In order to visualize the arrangement of the layers of pericardium, first observe the

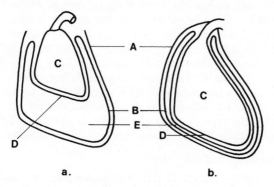

Fig. 107. Pericardium

fibrous pericardium (Fig. 107*a*) as an open bag (A) completely lined with serous pericardium (B) which does not open at the top. As the heart (C) develops, it enters the bag and pushes in front of it a layer of serous pericardium (D). The space (E) will be the **pericardial cavity**. The final relationship is shown in Fig. 107*b*. The fibrous pericardium (A) is lined with serous pericardium (B). A continuation of the serous pericardium (D) adheres to the heart muscle (C). The space between the two adjacent layers of serous pericardium is the pericardial cavity (E). Normally, this cavity contains a thin layer of fluid. In some diseases of the heart, the pericardial cavity contains excess fluid or blood. Since the outer fibrous bag is tough and unyielding, any excess fluid will interfere with the pumping action of the heart.

Parts of the Heart

The adult male heart weighs about 300 grams (about 11 oz.), and the adult female heart weighs about 250 grams (about 9 oz.). A front

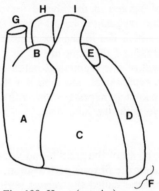

Fig. 108. Heart (anterior)

view of the heart is shown in Fig. 108. The **right atrium** (A) and its ear-like process, the **auricle** (B), form the right border of the heart. The **right ventricle** (C) takes up most of the anterior surface, while the **left ventricle** (D) forms most of the left border. The auricle of the **left atrium** (E) is just visible at the upper left corner. Most of the left atrium is seen from the posterior aspect. The **apex** of the heart (indicated by F) swings forward and upwards during contraction of the ventricles and hits the chest wall. This **apex beat** can usually be felt in the space below the fifth rib, about $3\frac{1}{2}$ inches from the midline. Three great vessels are seen at the upper border of the heart. They are the **superior vena cava** (G), the **aorta** (H) and the **pulmonary artery** (I).

Flow of Blood through the Heart

Fig. 109*a* is a diagram of the heart in **diastole**, or relaxation; Fig. 109*b* shows **systole**, or contraction of the ventricles. The heart consists of four chambers: left atrium, right atrium, left ventricle and right ventricle. The right and left atria are separated by the **interatrial septum**; the right and left ventricles by the **interventricular septum**. Unoxygenated venous blood enters the right atrium from the vena cava (A), during diastole. Oxygenated blood enters the left atrium at the same time from the pulmonary veins (B). Contraction of the right atrium forces the blood through the **tricuspid valve** (C) into the right ventricle. Contraction of the left atrium at the same time forces oxygenated blood into the left ventricle through the **mitral valve** (D). The tricuspid valve has three cusps; the mitral valve has two, which have the appearance of a bishop's mitre.

Contraction of the right ventricle forces blood through the **pulmonary valve** (E) into the pulmonary artery. Contraction of the left ventricle forces blood through the **aortic valve** into the aorta. The

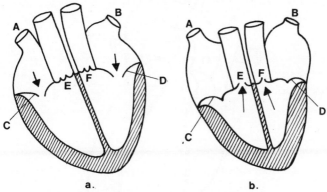

a.   b.

Fig. 109. Flow of blood through heart

valves which guard the openings of the arteries are known as the **semilunar valves**. Each valve consists of three little pockets facing into the artery, and each pocket is shaped like a half-moon. During diastole, the valves between the atria and the ventricles are open and the semilunar valves are closed. During systole, the semilunar valves are forced open by the pressure of blood from the ventricles, and the tricuspid and mitral valves are forced shut.

Before birth, the lungs are not functioning and so the pulmonary circulation is much less important. In the foetal heart there is an opening in the interatrial septum which permits blood to pass directly from the right atrium into the left atrium without flowing through the lungs at all. This opening, the **foramen ovale**, usually closes after birth, leaving an oval depression called the **fossa ovale**. In about 10 per cent of the population, the foramen ovale remains open, which is one type of congenital heart defect. It is not necessarily serious, as occasionally the opening is so small that blood does not pass through it.

Main Blood-vessels of the Heart

The blood-vessels associated with the heart are shown in Fig. 110. The superior vena cava (A) and the inferior vena cava (B) empty into the right atrium (C). Blood from the right ventricle (D) is pumped into the **pulmonary trunk** (E), which divides into the right and left pulmonary arteries just above the heart. The left ventricle (F) pumps blood into the **ascending aorta** (G). From the **arch of the aorta** (H), three large arteries arise to supply the head and neck and the upper limbs. On the right, the **innominate artery** (I) divides into the **common carotid**

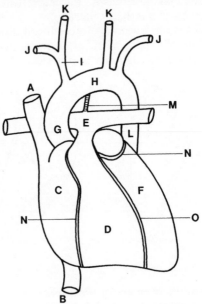

Fig. 110. Heart and main vessels

artery (K) and the **right subclavian artery** (J). The **left common carotid artery** (K) and the **left subclavian artery** (J) arise directly from the arch of the aorta. The carotid arteries supply the head and neck, and the subclavian arteries supply the upper limbs. The **descending aorta** (L) continues behind the heart and descends through the thoracic cavity in the posterior mediastinum.

A cordlike structure (M), the **ligamentum arteriosum,** is seen connecting the arch of the aorta with the pulmonary trunk. In the foetus, an actual communication, or shunt, exists between these two vessels, and blood is shunted from the pulmonary circulation to the aorta, as the lungs are not functioning. After birth this shunt, called the **ductus arteriosus,** closes off, becoming a cordlike structure. Failure to close results in another congenital defect, **patent ductus arteriosus.**

If left uncorrected it causes overwork, and eventual failure, of the left ventricle and excessive pressure in the pulmonary circulation. Fortunately, it is a relatively easy defect to correct.

The groove (N) indicates the line of separation between the atria and the ventricles and is called the **atrio-ventricular sulcus**. The groove (O) on the anterior surface of the heart indicates the site of the **inter-ventricular septum.**

The **coronary arteries** (Fig. 111) arise from the aorta, just above the aortic valve. The **right coronary artery** (A) runs in the atrio-ventricular sulcus and turns around the right border of the heart to lie on its posterior aspect (C). The **left coronary artery** (B) soon divides into a

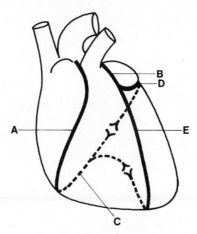

Fig. 111. Coronary arteries

circumflex branch (D) which runs in the left atrio-ventricular sulcus to supply the posterior surface. The other branch (E), the inter-ventricular artery, descends to supply the anterior surfaces of both ventricles.

Blood returning from the heart is collected into veins which follow the course of the arteries and eventually empty into one vessel, the **coronary sinus**. The coronary sinus itself empties into the right atrium. Because there is not much collateral circulation in the heart, it follows that if one coronary artery is blocked by a clot of blood (thrombosis) the muscle supplied by that artery is starved of blood and dies. If a major coronary artery is blocked, the outlook is serious.

Lymphatic System

The lymphatic vessels carry lymph from all the tissues of the body, except the brain and spinal cord. The vessels unite with one another to form large tubes until there are finally two main channels. The

thoracic duct ascends through the thorax in front of the vertebrae and ends by joining the left subclavian vein in the root of the neck. Blood does not enter the duct because the junction is protected by a valve. The **right lymphatic duct** is formed from all the lymphatics coming from the right arm and the right side of the head and neck. It empties into the right subclavian vein.

The Diaphragm

The diaphragm is the dome-shaped septum separating the thoracic cavity from the abdominal cavity. It has a peripheral muscle part and a

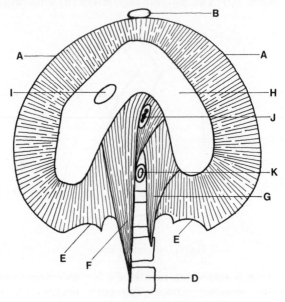

Fig. 112. Diaphragm (abdominal aspect)

central tendinous part called the **aponeurosis**. In Fig. 112 the diaphragm is shown from its abdominal or concave surface. The muscular fibres arise from the lower six ribs (A) and the sternum (B). This is called the **costal part**. Other muscular fibres arise from the vertebral bodies (D) and the **arcuate ligaments** (E). This is called the **vertebral part**. The **right crus** (F) is a strong fibrous straplike process which is attached to the front of the bodies of the upper three lumbar vertebrae. The **left crus** (G) is only attached to the upper two vertebrae. The arcuate ligaments are a series of fibrous arches attached to the fascia covering the muscles of the back. The muscular fibres are inserted into the central aponeurosis (H).

There are three main openings in the diaphragm. The inferior cava

(I) pierces the aponeurosis. The oesophagus (J) descends in the posterior mediastinum and pierces the muscular part of the diaphragm in relation to the right crus. The descending aorta (K) passes under the arch formed by the medial borders of the two crura (plural of *crus*) and lies anterior to the vertebral bodies.

THE ABDOMEN

The muscles of the anterior wall of the abdomen and the lower thorax are shown in Fig. 113. The most superficial muscle in the space between the ribs is the **external intercostal muscle** (A), shown here only

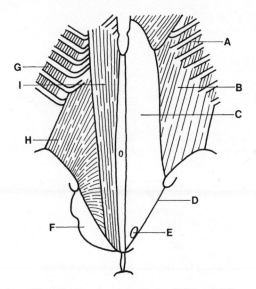

Fig. 113. Anterior abdominal wall (superficial)

on the left side of the body. This muscle elevates the ribs and so increases the capacity of the thoracic cavity during inspiration (breathing in). On the anterior abdominal wall, the **external oblique muscle** (B) is the most superficial; its fibres run downwards and medially. The muscular fibres are inserted into a sheet of fibrous tissue called an **aponeurosis** (C). This aponeurosis extends to the midline, where its fibres are continuous with the right external oblique aponeurosis (not shown). The inferior border of the aponeurosis is folded under to form the **inguinal ligament** (D). This is attached laterally to the anterior superior spine of the ilium and medially to the pubic bone, and it bridges over an opening (F) between the anterior part of the pelvis and the thigh, shown on the right side of the diagram. The opening in the

aponeurosis itself (E) is the **external inguinal ring**. The spermatic
cord passes through this opening and will be described later.

On the right side of the body the superficial muscle layer has been
removed. The fibres of the **internal intercostal muscle** (G) run upwards
and medially. The muscle fibres of the corresponding layer on the
abdominal wall run in the same direction. This is the **internal oblique
muscle** (H), which is inserted into a medial aponeurosis. The **rectus
abdominis muscle** (I) runs from the thorax to the front of the pelvis.

The third and deepest layer of muscle is shown in Fig. 114. The
innermost intercostal muscle fibres (A) run obliquely, but the **trans-**

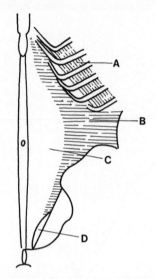

Fig. 114. Anterior abdominal wall (deep)

versus abdominis muscle fibres (B) run transversely, as the name
implies. The greater part of the aponeurosis of the transversus
abdominis (C) passes deep to the rectus abdominis muscle. A layer of
fascia lies under the third muscle layer in both the thorax and
abdomen. In the thorax it is called the endothoracic fascia, and in the
abdomen it is called the transversalis fascia. The opening (D) in the
transversalis fascia indicates the site of the **internal inguinal ring**,
through which passes the spermatic cord. Deep to the fascial layer,
parietal pleura lines the thoracic cavity and parietal peritoneum lines
the abdominal cavity.

A transverse section through the anterior abdominal wall (Fig.
115) illustrates the way in which the layers of aponeurosis form a
sheath round the rectus abdominis muscles (A). The external oblique
aponeurosis (B), just under the skin, fuses with the anterior sheet of the

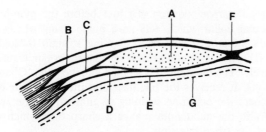

Fig. 115. Anterior abdominal wall (transverse section)

internal oblique aponeurosis (C) to form the **anterior rectus sheath**. The posterior sheet of the internal oblique aponeurosis fuses with the transversus aponeurosis (D) to form the **posterior rectus sheath**. Both sheaths join together at the midline to become continuous with their opposite numbers and to form a strong fibrous layer called the **linea alba** (F). The transversalis fascia (E) is a separate layer, and the peritoneum (G) lies deep to it.

The Stomach and Duodenum

In Fig. 116 the oesophagus (A) joins the **stomach** at the **cardia** (B) after it has passed through the diaphragm. The stomach projects upwards to the left to form the **fundus** (C). The body of the stomach

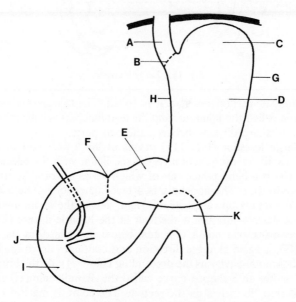

Fig. 116. Stomach and duodenum

(D) leads to a narrow portion (E) just before the **pyloric sphincter** (F), a muscular collar which acts as a purse string when contracted, and prevents the stomach contents passing straight through into the **duodenum** (I). The **lesser** and **greater curvatures** are indicated at H and G, respectively. The duodenum is shaped like the letter 'C' and is about 10 inches long. It receives the **bile duct** and the **pancreatic duct** which arrive together; the common opening is guarded by the **sphincter of Oddi** (J). At K the duodenum makes a sharp turn which marks the beginning of the small intestine.

The Intestines

The length of the **small intestine** varies; the average length is about 20 feet. It is suspended from the posterior abdominal wall in a fan-

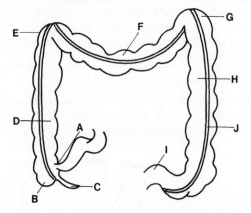

Fig. 117. Large intestine

shaped fold of peritoneum and it is coiled. The first part of the small intestine is called the **jejunum**, and the terminal part is called the **ileum**. There is no sharp division between the two parts.

The **large intestine** (Fig. 117) begins in the lower right of the abdominal cavity at the junction (A) of the ileum with the **caecum**. The caecum (B) is a blind pouch out of which arises a wormlike structure, the **appendix** (C). The appendix is a few inches long. The **ascending colon** (D) extends upwards on the right side of the abdominal cavity and turns sharply just below the liver at the **hepatic flexure** (E) to become the **transverse colon** (F). Just below the spleen, at the **splenic flexure** (G), it again changes direction and continues down the left side of the abdominal cavity as the **descending colon** (H). At the brim of the pelvis it makes an S-shaped curve called the **sigmoid colon** (I) and then descends into the pelvis as the **rectum**. The colon is marked by three longitudinal bands of muscle which commence at the base of the appendix and end at the recto-sigmoid junction. These bands, known as the **taenia coli** (J), are shorter than the gut to which they are attached, so they pull the walls of the colon into their typical **sacculations**.

Liver, Gall Bladder, Pancreas and Spleen

The **liver** is the largest organ in the body, weighing about 3 pounds in the adult male. It lies below the diaphragm in the upper right quadrant of the abdomen. Fig. 118 is the anterior view of the liver with the

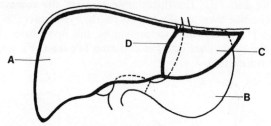

Fig. 118. Liver (anterior)

position of the stomach (B) behind it indicated by a dotted line. The large **right lobe** (A) is separated from the **left lobe** (C) by the **falciform ligament** (D) which is a sickle-shaped ligament connecting the liver to the diaphragm and the anterior wall of the abdomen. The falciform ligament extends as far as the umbilicus, and in the foetus it contains the blood vessels of the umbilical cord.

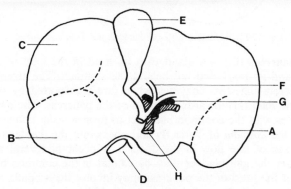

Fig. 119. Liver (inferior)

The inferior surface of the liver (Fig. 119) shows the impressions produced by the adjacent structures: the **gastric impression** (A) produced by the stomach, the **renal impression** (B) produced by the right kidney, and the **colic impression** (C) produced by the hepatic flexure of the colon. The inferior vena cava (D) is closely related to the posterior surface of the liver. The **gall bladder** (E) is a sac lying on the inferior surface of the liver, which receives bile from the **hepatic ducts** (F) and stores it before discharging it into the duodenum. The blood-vessels of the liver enter where the hepatic ducts leave. The **hepatic artery** (G) brings oxygenated blood to the tissues of the liver. The **portal vein** (H) brings blood rich in digested food from the intestines, and **iron** from the spleen.

The bile produced in the liver is drained from the left lobe by the left hepatic duct (A in Fig. 120) and from the right lobe of the liver by the right hepatic duct (B). The ducts unite to form the **common hepatic duct** (C). The bile then usually travels via the **cystic duct** (D) to the gall bladder (E). However, when we are eating, bile flows from the hepatic ducts and the cystic duct into the **common bile duct** (F), which passes behind the duodenum.

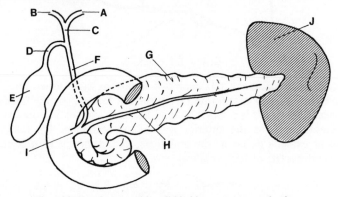

Fig. 120. Duodenum, with gall-bladder, pancreas and spleen

The **pancreas** (G) is a gland with its head in the loop of the duo-denum, and its body and tail behind the stomach. The **alveoli** of the gland secrete digestive enzymes which are transported by the main **pancreatic duct** (H) towards the head of the pancreas. The pancreatic duct unites with the common bile duct to form a wider channel which is called the **ampulla of Vater** (I) and opens into the duodenum.

Blockage of bile flow in the ducts may occur in several areas. A stone from the gall bladder may be caught in the common bile duct. Cancer of the head of the pancreas may involve the ampulla of Vater. If the flow of bile is prevented, the level of bile pigments in the blood increases, producing **jaundice.**

The **spleen** (J) is a soft, friable organ enclosed by a tough coat of fibrous elastic and muscular tissue. It lies on the left side of the stomach, and is closely related to the tail of the pancreas. Its structure is similar to that of a lymph node, but instead of lymph it contains blood. It destroys worn-out red blood cells, acts as a reservoir for blood cells of all kinds, and is the largest lymphoid organ in the body. In certain infectious diseases it enlarges, and it is probably concerned with the development of antibodies. However, the spleen can be removed quite safely, so its functions are not essential to life. It has to be removed when injured, as it bleeds excessively if the capsule is torn, which may happen in cases of violent blunt injury to the abdomen.

Blood Supply to the Abdomen

The aorta emerges from under the arch of the diaphragm and almost immediately gives rise to the **coeliac axis**, or coeliac trunk, which is illustrated in Fig. 121. The coeliac trunk (A) is usually very short;

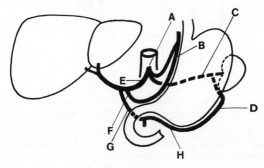

Fig. 121. Coeliac axis

it divides into three branches. The first branch is the **left gastric artery** (B) which runs along the lesser curvature of the stomach. The second branch, the **splenic artery** (C), runs behind the stomach to the spleen and gives off the **left gastro-epiploic artery** (D) which runs along the greater curvature of the stomach.

The third branch of the coeliac trunk is the **hepatic artery** (E) which gives off two branches. The **right gastric artery** (F) runs along the lesser curvature of the stomach to anastomose with the left gastric artery. The **gastro-duodenal artery** (G) sends branches to the duodenum and pancreas and also gives rise to the **right gastro-epiploic artery** (H) which travels along the greater curvature of the stomach to anastomose with the left gastro-epiploic artery. The hepatic artery continues to the liver and gives branches to the right and left lobes, and also a branch to the gall bladder.

The blood supply to the intestines is shown in Fig. 122. The **superior mesenteric artery** (A) arises from the aorta just below the coeliac axis. It gives off many branches: to the small intestine (B); the **ileo-colic artery** (C) supplying terminal ileum, caecum and appendix; the **right colic artery** (D) supplying the ascending colon; and the **middle colic artery** (E) supplying the transverse colon. The inferior **mesenteric artery** (F) arises from the aorta and has three branches: the **left**

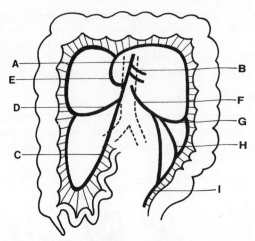

Fig. 122. Arteries of the intestines

colic artery (G) which supplies the descending colon; the **sigmoid artery** (H) which supplies the sigmoid colon; and the **superior haemorrhoidal artery** (I) which supplies the rectum. Each branch of these arteries anastomoses freely with its neighbour, so there is a continuous arcade of arteries along the whole length of the gastro-intestinal tract.

The other branches of the aorta will be described later, in connection with the organs they supply.

The Portal System of Veins

Venous blood from the gastro-intestinal tract is rich in digested food which must be processed further in the liver. The veins which drain the abdominal part of the digestive tract, the pancreas and the spleen, unite to form the portal vein, which empties into the liver. This network of veins is called the **portal system**. The distal veins correspond to the arteries described above and they drain into the veins illustrated in Fig. 123. The **inferior mesenteric vein** (A) enters the **splenic vein** (B) behind the pancreas. The **gastric vein** (C) joins the splenic vein, which then unites with the **superior mesenteric vein** (D) to form the portal vein (E). The portal vein enters the liver and breaks up into a network of thin-walled blood channels called **sinusoids**. From the sinusoids, the blood is drained into the right and left **hepatic veins** (F) which enter the inferior vena cava (G) near its termination in the right atrium.

Normally, portal venous blood flows along the system of vessels as described. However, if this pathway is blocked in any way, e.g. in cirrhosis of the liver, the portal venous pressure rises and the blood is forced through collateral channels. There are several areas where the

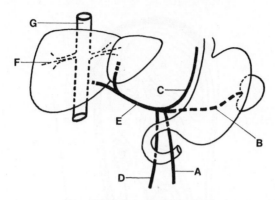

Fig. 123. Portal system of veins

portal system of veins anastomoses with the systemic system, i.e. with the veins which drain directly into the vena cavae (cavae is the plural of cava). One area is at the base of the oesophagus and another is at the lower end of the rectum. The veins in these areas are surrounded by loose connective tissue and if the pressure rises they become dilated and tortuous, i.e. varicose. Varicose veins in the oesophagus are liable to burst and give rise to the **haematemesis** (vomiting of blood) that can occur in portal hypertension.

The Peritoneum

The abdominal cavity shown in Fig. 124 is lined with a thin, smooth, shiny layer of tissue called peritoneum. The peritoneum that lines the wall of the cavity is called parietal peritoneum. Note the layer of parietal peritoneum on the anterior abdominal wall (N) and follow it upwards to the inferior surface of the diaphragm (I). The peritoneum is reflected off the diaphragm on to the liver (A), which it almost completely surrounds. This layer is called visceral peritoneum because it covers a **viscus**, or organ. The same layer is reflected off the liver to the stomach (B) and forms part of a fold of peritoneum hanging from the stomach, called the **greater omentum** (L). Return to the parietal peritoneum on the anterior abdominal wall and follow it down into the pelvis, where it is reflected on to the bladder (F) and the rectum (G) and the posterior abdominal wall. The more detailed arrangements of the pelvic peritoneum will be described later in connection with the sexual organs.

The intestines are covered by peritoneum in a similar manner to that described for the lungs. The double layer of peritoneum suspending the small intestine (E) from the posterior abdominal wall is called the **mesentery** (M). Arteries from the aorta (H) run between the two layers to reach the intestine, along with veins, nerves and lymphatic vessels.

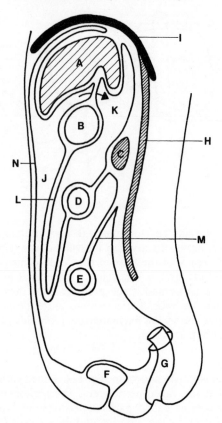

Fig. 124. Peritoneum

The transverse colon (D) has a similar mesentery, also conveying blood-vessels, nerves and lymphatics. The pancreas (C) is covered by the parietal peritoneum, but has no mesentery. Follow this layer up to the inferior surface of the diaphragm where it is reflected off the diaphragm on to the liver, then down to the posterior surface of the stomach. This layer continues downwards to complete the greater omentum.

The spaces J and K are called the **peritoneal cavity**. This cavity is very small because the organs covered with visceral peritoneum fill the abdominal cavity and are everywhere in contact with the parietal peritoneum. A small amount of fluid in the cavity lubricates the sliding movements of the viscera. In some conditions, excess fluid, called **ascites**, is produced in the peritoneal cavity and this can be drawn off by the insertion of a wide-bore needle through the anterior abdominal wall. The long mesentery of the small intestine may become twisted, shutting off the blood supply to the wall of the intestine. This condition is called **volvulus**.

The double peritoneal layer suspending the stomach from the liver is called the **lesser omentum**. It contains the bile duct, the hepatic artery, the portal vein and lymphatic vessels. The lymphatics can carry cancer cells from a growth in the stomach to the liver, where they form secondary growths, or **metastases**. The stomach and the lesser omentum divide the peritoneal cavity into two areas. The **greater sac** (J) communicates with the **lesser sac** (K) by means of a channel through the peritoneum, indicated by the arrow. This is the **foramen of Winslow**.

Infection in the peritoneal cavity, or **peritonitis**, is very serious because toxic bacterial products can be absorbed rapidly through the peritoneum into the blood stream. The speed of absorption through a membrane is related to the surface area of the membrane, and the area of the peritoneum is approximately equal to the individual's skin area.

The Urinary Tract: The Kidneys

The relationship of the **right kidney** to the posterior abdominal wall is shown in Fig. 125. The diaphragm (A) forms the upper part of the

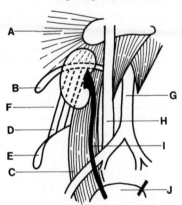

Fig. 125. Right kidney and ureter

posterior abdominal wall before arching forwards to form the roof of the abdominal cavity. Below the twelfth rib (B) three muscles lie behind the kidney: the psoas (C) which unites with the iliacus in the pelvis to form the iliopsoas muscle, the **quadratus lumborum** (D) which runs from the twelfth rib to the iliac crest (E), and the transversus abdominis muscle (F) which sweeps round to form part of the lateral and anterior abdominal walls. The aorta (G) emerges under the arch of the diaphragm and runs down anterior to the vertebrae. At the level of the fourth lumbar vertebra it divides into the **right** and **left common iliac arteries**. The inferior vena cava (H) lies to the right of the aorta. The **ureter** (I) is about 10 inches long and runs down on the medial border of the psoas muscle, lateral to the vena cava, crossing in front of the bifurcation of the iliac artery, before reaching the **bladder** (J).

The posterior relationships of the left kidney are the same. Both kidneys lie behind the peritoneum; the right kidney is a little lower than the left because it is pushed downwards by the larger right lobe of the liver.

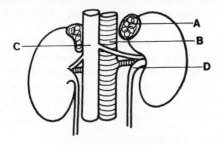

Fig. 126. Kidneys (anterior)

The blood supply of the kidneys is shown in Fig. 126. The upper end of each kidney is capped by the **suprarenal gland** (A). Arising from the aorta (B), the **renal arteries** supply the kidneys with blood at high pressure. The renal veins return the blood to the inferior vena cava (C). The blood vessels enter a vertical slit, the **hilum**, on the medial aspect of each kidney. Emerging from the hilum is the **pelvis** of the ureter (D) and also the nerves and lymphatic vessels.

The cut surface of the kidney is shown in Fig. 127. It is covered by a layer of fascia (A) called **renal fascia**. This is separated by a layer of

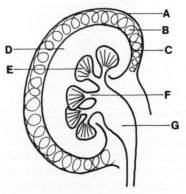

Fig. 127. Sagittal section of kidney

fat, the **perinephric fat** (B), from the true capsule of the kidney (C). The outer layer of kidney substance is called the **cortex** (D) and it contains the filtering system. The inner layer is called the **medulla** and it consists of the renal **pyramids** (E) which are made up of small tubules. The urine is finally emptied into a **minor calyx** (F), and several calyces (plural of calyx) join together to form the pelvis of the ureter.

Kidney stones are sometimes formed in the renal pelvis. If the stone is small enough it may be pushed along the ureter, resulting in severe pain. Stones may be held up at three constrictions of the ureter: its junction with the renal pelvis, the point where it bends over the brim of the true pelvis, and the point where it enters the muscular wall of the bladder.

The Bladder and Urethra

The bladder is very elastic. It can hold about a pint of urine, but is usually emptied when the volume of urine is about half a pint. It lies behind the pubic symphysis and is covered by a layer of peritoneum. The ureters empty through the posterior wall. The male urethra is about 8 inches long and the female urethra about $1\frac{1}{2}$ inches. They will be described separately in relation to the genital organs.

Male Genital Organs: The Prostate

The prostate is an organ composed of fibrous, muscular and glandular tissue. It is about the size and shape of a chestnut and it surrounds the urethra at the base of the bladder. Fig. 128 is a posterior

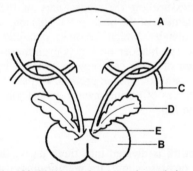

Fig. 128. Bladder and prostate (posterior)

view of the bladder (A) and the prostate (B). The **vas deferens** (C) and the **seminal vesicle** (D) unite to form the **ejaculatory duct** (E) which traverses the prostate to empty into the urethra. The vas deferens carries **semen** from the testis. The prostate and the seminal vesicles produce part of the seminal fluid. The prostatic secretion reaches the prostatic urethra through about a dozen little ducts which empty into the posterior aspect of the urethra.

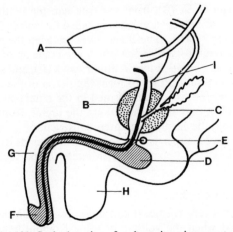

Fig. 129. Sagittal section of male genito-urinary system

The sagittal section of the same structures is shown in Fig. 129. The bladder (A) is shown resting on top of the prostate (B). The first part of the urethra traverses the prostate, and receives the ejaculatory duct (C) as well as the prostatic ducts (not shown). The posterior part of the prostate is divided by the ejaculatory duct into two lobes: the **median lobe** above the duct, and the **posterior lobe** below and behind the duct.

It can be seen from Fig. 129 that enlargement of the median lobe is likely to put pressure on the upper part of the urethra, where it kinks to enter the neck of the bladder (I). This may obstruct the flow of urine from the bladder, which becomes painfully distended. As the bladder enlarges, the bladder neck gets even more obstructed and the resulting **retention** of urine can only be relieved by the passage of a hollow tube, or **catheter**, up the urethra. Hypertrophy of the prostate is part of the process of normal ageing, but luckily it does not always affect the median lobe.

The Urethra

The male urethra is divided into three parts: the **prostatic urethra** enclosed by the prostate gland; the **membranous urethra** which pierces the perineal membrane; and the **spongy urethra**, the final 6-inch section which traverses the **corpus spongiosum** of the **penis**. Just as it enters the corpus spongiosum, the urethra makes a sharp turn and then runs upwards and forwards to lie below the pubic symphysis, then it turns downwards and forwards when the penis is flaccid. The corpus spongiosum is composed of spongy, erectile tissue and it surrounds the urethra in its course through the penis. (Erection of the penis occurs when the spaces inside the corpus spongiosum become filled with blood.)

The area of the right-angled turn is called the **bulb** of the urethra (D), and the enlargement of the corpus spongiosum at the tip of the penis is called the **glans penis** (F). The paired **bulbo-urethral glands** (E) open into the urethra at the bulb.

The penis also contains **corpora cavernosa** (G) which lie on top of the corpus spongiosum and are attached to the conjoined rami of the hip bone. The root of the penis will be described later, along with the perineum. The **scrotum** (H) is a pouch which contains the **testes** (plural of *testis*) and their coverings.

The Testis and Epididymis

In the foetus, the testes first appear near the developing kidneys on the posterior abdominal wall behind the parietal peritoneum. As the testes enlarge, they begin to descend towards the scrotum and this

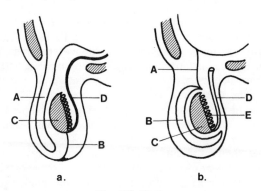

a.   b.

Fig. 130. Testis

process explains some of the features of their anatomy. Fig. 130 is a sagittal section through the right testis, showing an early stage in Fig. 130*a* and the final stage in Fig. 130*b*.

In the third month of foetal life, the testis reaches the pelvis and at the same time a sac-like projection of the peritoneum begins to push through the abdominal wall and descend into the scrotum. This is called the **processus vaginalis** (A). A fibro-muscular strand, the **gubernaculum testis** (B), extends from the testis along its path of descent into the scrotum behind the processus vaginalis. The testis (C) and its associated vas deferens (D) usually reach the scrotum at birth. Shortly afterwards, the processus vaginalis closes off, and is no longer connected with the abdominal cavity.

The normal adult testis is shown in Fig. 130*b*. The processus vaginalis (A) has closed off and that part of it surrounding the testis forms a double-walled sac, the **tunica vaginalis** (B). The semen and fluid formed in the testis (C) pass into a coiled tube, the **epididymis** (E), which lies on the posterior wall of the testis and joins the vas deferens (D) which passes up through the scrotum.

The blood supply, nerves and lymphatic drainage of the testis are all derived from the renal area where the testis originates. Secondaries from cancerous growth in the testis will consequently pass up lymphatic vessels to the posterior wall of the abdomen and may even spread as far as the neck. Pain felt in the kidney or ureter may be referred down to the scrotum; conversely, testicular pain may be felt in the loin. Sometimes the processus vaginalis fails to close completely and leaves a passage from the abdominal cavity to the scrotum down which coils of intestine may pass. Such a condition, called an **indirect inguinal hernia**, can be repaired by pushing the intestine back up again the tying off the processus vaginalis.

Inguinal Canal

The inguinal canal is the oblique pathway through the anterior abdominal wall, which is traversed by the testis in its descent to the scrotum. The **spermatic cord** contains the vas deferens, the blood-vessels, lymphatics and nerves, which are associated with the testis. It passes from the testis through the inguinal canal into the abdominal cavity. The canal has a superficial opening called the external inguinal ring in the aponeurosis of the external oblique muscle. The other end is a deeper opening in the transversalis fascia called the internal inguinal ring.

Fig. 131*a* shows the superficial structures of the inguinal region. The **inguinal ligament** (A) is the lower border of the aponeurosis of the external oblique muscle (B) and extends from the anterior superior iliac spine (C) to the pubic bone (D). The femoral artery (E) and femoral vein (F) pass under the ligament to reach the thigh. They are quite near the skin in this region. The spermatic cord (G) passes through the external inguinal ring (H) to reach the scrotum.

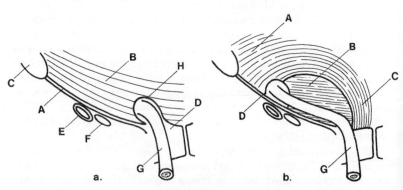

Fig. 131. Inguinal region

Fig. 131*b* shows the same area after removal of the external oblique muscle. The internal oblique muscle (A) joins the transversus muscle to form the **conjoint tendon** (C) which is inserted into the pubic bone. Deep to the muscle is the transversalis fascia (B) which forms the inner layer of the abdominal wall. The internal inguinal ring (D) represents the point where the processus vaginalis originally pushed through the transversalis fascia and then turned downwards and medially to reach the scrotum. This is the path followed by the spermatic cord.

The Spermatic Cord

The spermatic cord acquires a coat from each layer of the abdominal wall. Fig. 132 is a cross section of the cord which identifies the various layers. The inner layer is derived from the transversalis fascia and is called the **internal spermatic fascia** (A). The next layer is derived from the internal oblique muscle and consists of muscle fibres, the

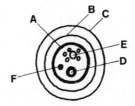

Fig. 132. Transverse section of spermatic cord

cremaster muscle (B). The cremaster muscle can contract and pull the testis towards the inguinal canal. (This happens more easily in childhood and must be remembered when a small boy appears to have an empty scrotum. It may indicate that the testes have failed to descend or that they are simply pulled up by the cremaster muscle, a condition known as **retractile testis**.) The outer layer is derived from the aponeurosis of the external oblique muscle, and is called the **external fascia** (C).

The spermatic cord contains the vas deferens (D) and the **testicular artery** (E), surrounded by a plexus of veins called the **pampiniform plexus**. If these veins become varicose, the condition is called **varicocoele**. Lymph vessels and nerves are also associated with the artery. The fibrous remains of the processus vaginalis (F) are also found in the cord.

Female Genital Organs: The Uterus

The uterus and its associated organs are shown in Fig. 133, which is a sagittal section through the female pelvis. The **uterus** (A) is pear-shaped and about 3 inches long. The **Fallopian tube** (B) runs from the upper lateral aspect of the uterus towards the **ovary** (C). The

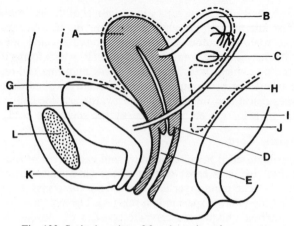

Fig. 133. Sagittal section of female genito-urinary system

cervix (D) is the lower end of the uterus and it communicates with the **vagina** (E) which opens to the exterior just behind the **urethra** (K). The bladder (F) lies in front of the uterus and is separated from it by a pouch of peritoneum (G) (which is continuous with the parietal peritoneum of the abdominal wall). The ureter (H) is closely related to the lateral wall of the cervix. The rectum (I) is separated from the posterior wall of the uterus by a fold of peritoneum called the **pouch of Douglas** (J).

Enlargement of the pregnant uterus presses on the bladder and decreases its capacity. Frequency of **micturition** (passage of urine) is a feature of pregnancy because of this relationship. The urethra is near the anterior wall of the vagina. During childbirth the baby's head presses the urethra against the pubic symphysis (L); this may injure the urethra and make micturition painful for a few days following delivery.

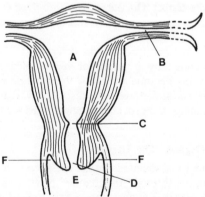

Fig. 134. Coronal section of uterus

The uterus is bisected in Fig. 134 showing its cavity (A) which is continuous with the hollow Fallopian tubes (B). The **cervical canal** begins at the **internal os** (C) and opens into the vagina (E) at the **external os** (D). Thus there is a direct connection between the outside world and the peritoneal cavity. This is of practical importance in cases of infection of the vagina, e.g. **gonorrhoea**, which can extend upwards and result in peritonitis. The cervix projects into the vagina, and the upper part of the vagina which surrounds the cervix is called the **fornix** (F).

The normal relationship of the uterus to the vagina is shown in Fig. 135a. The vagina is about $3\frac{1}{2}$ inches long (it distends during intercourse and delivery) and is inclined backwards. The cervix projects through

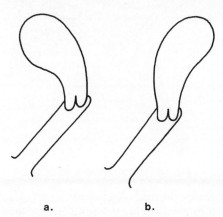

a.   b.

Fig. 135. Sagittal section of uterus and vagina

the anterior wall at a right-angle to the long axis of the vagina. This is known as **anteversion**. The uterus is normally bent forward in itself; this angle is known as **anteflexion**. An abnormal position is shown in Fig. 135b, where the uterus is bent back on itself (**retroflexion**) and the cervix is no longer at right-angles to the long axis of the vagina (**retroversion**). This position means that it is easier for the uterus to slide down the vagina, an abnormality known as **prolapse**.

The external os of the uterus is not normally in line with the vagina, which is an anatomical fact of great importance in the passage of instruments into the uterus. The close relationship of the pouch of Douglas to the posterior fornix of the vagina has resulted in tragedy for some victims of amateur abortionists. An instrument, e.g. a knitting needle, passed up the vagina can miss the cervix and get pushed through the posterior fornix instead. This carries infection into the peritoneal cavity with sometimes fatal results.

The Ovary

Fig. 136 shows the posterior aspect of the uterus and its relations on the right side. The ovary (A) is an oval structure about $1\frac{1}{2}$ inches long. Like the testis, it develops in the renal area on the posterior abdominal wall and then descends, only not so far. Its blood-vessels, lymphatics and nerves follow it down from its origin in a fold of peritoneum called the **infundibulo-pelvic ligament** (C). The **ovarian ligament** (D) is a strand of fibrous tissue which connects the ovary with the uterus (B) just below the origin of the Fallopian tube (E).

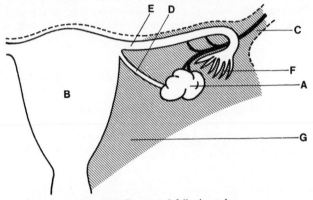

Fig. 136. Ovary and fallopian tube

The ovary releases normally one **ovum** every month, and this is usually swept up by the **fimbriae** (F) at the end of the Fallopian tube. The ovum passes down to the uterus and, if fertilized, is implanted in the uterine wall. Very rarely, a fertilized ovum falls into the peritoneal cavity or remains in the Fallopian tube. This is known as **ectopic pregnancy**; it may come to term in the abdominal cavity, but it always ruptures in the case of tubal pregnancy. In either case, an operation is required to remove the products of conception and usually the tube also. The ovary lies on the posterior wall of the broad ligament (G).

The Broad Ligament

Fig. 137 shows a sagittal section through the uterus (B) to illustrate the relationships of its lateral wall. The broad ligament is a fold of peritoneum which covers the uterus and the related structures on its lateral aspect. The dotted line A shows the attachment of the broad ligament to the uterus. Included in the fold are the Fallopian tube (C), the ovarian ligament (D), the **round ligament** (E), and the **uterine artery** (F) which hooks over the ureter (H). The round ligament passes from the uterus through the inguinal canal to end in the **labium majus**, which is equivalent to the male scrotum in terms of anatomical development.

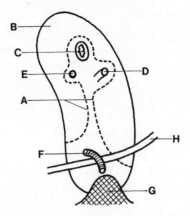

Fig. 137. Uterine attachment of broad ligament

The round ligament and the ovarian ligament are together the equivalent of the gubernaculum testis. The broad ligament, being a fold of peritoneum, does not give much support to the uterus. The strong fibrous **cardinal ligament** (G) is an important lateral support of the uterus.

Supports of the Uterus

By removing the bulk of the pelvic organs, a view of the pelvic ligaments is revealed as in Fig. 138. The **cardinal ligaments** (A) extend from the cervix (B) to the side walls of the pelvis. The ureters pierce them to reach the bladder. The **utero-sacral ligaments** (C) curve backwards from the cervix to the periosteum in front of the sacro-iliac joints and the third part of the sacrum (D). They form the lateral boundaries of the pouch of Douglas (E), and encircle the rectum (F). The **pubo-cervical fascia** (G) extends forwards from the cardinal ligaments and acts as a sling for the bladder (H) before becoming attached to the pubis (I).

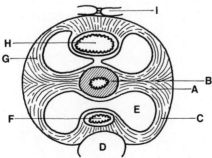

Fig. 138. Supports of the uterus

THE PELVIC FLOOR AND PERINEUM

Look back at Figs. 66, 67 and 68 and revise the anatomy of the bony walls of the pelvis. The canal formed by the bony and ligamentous walls of the pelvis has a floor composed of muscles and sheets of fascia. This floor, called the **pelvic diaphragm**, is pierced by the urethra, the rectum, and the vagina in the female. Fig. 139 shows the

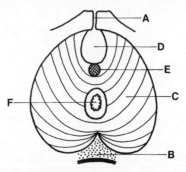

Fig. 139. Pelvic diaphragm

pelvic diaphragm as viewed from above. The anterior part is the pubic symphysis (A), and the muscle fibres sweep round to be attached to the side walls of the pelvis and the sacrum (B) posteriorly. The **levator ani** (C) is the largest and most important of the muscles of the pelvic floor. It forms a sling round the prostate, or vagina (D), and some fibres are inserted into the **perineal body** (E), a fibro-muscular mass which strengthens the midline of the pelvic floor. The posterior part of the muscle forms a sling round the rectum (F).

Fig. 140 is a sagittal section through the pelvis looking into the right half of the pelvic cavity. The levator ani (A) is shown sweeping down from its origin on the side wall of the pelvis from the pubic symphysis

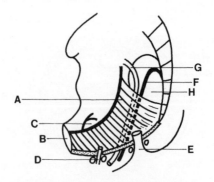

Fig. 140. Sagittal section of pelvic cavity

(B); it is supported on a fibrous arch over the obturator foramen (C). In the midline, the levator ani is pierced by the urethra (D) and the rectum (E). The structures under the pelvic diaphragm are in the **perineum**. The **pudendal nerve** (F) and the **pudendal artery** (G) arise within the pelvic cavity. They skirt the pelvic diaphragm by passing through the greater sciatic foramen (H) and then turning through the lesser sciatic foramen under the levator ani. The pudendal nerve branches to supply the structures of the perineum, which include the penis and the anal canal. The branches of the artery follow the nerves.

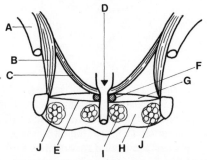

Fig. 141. Coronal section of pelvic floor and perineum

A frontal section through the pelvis and perineum (Fig. 141) shows the same structures, in a plane at right-angles to Fig. 140. Lining the bony wall of the pelvis (A) is the obturator internus muscle (B). The levator ani (C) arises from the fascial covering of the obturator internus and sweeps down to support the base of the bladder (D). The **uro-genital diaphragm** (E) is the floor of the anterior part of the perineum. It is pierced by the membranous part of the urethra. The **external sphincter** of the urethra is a collar of voluntary muscle fibres around the membranous part of the urethra. It is enclosed in the **deep perineal pouch** (G), which also contains the **deep transverse perineal muscles**. Inferior to the uro-genital diaphragm is the **superficial perineal pouch** (H), which contains the bulb of the urethra comprising the corpus spongiosum (I) and the crura of the corpora cavernosum (J). Urine from a ruptured urethra collects in the superficial perineal pouch.

The structures described so far are common to both sexes, but the other structures of the anterior perineum will have to be described separately as they include the genital organs.

Male Perineum

Fig. 142 shows the inferior surface of the male perineum. The coccyx (A) and the ischial tuberosities (B) are bony landmarks which

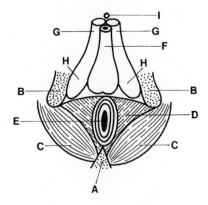

Fig. 142. Male perineum

you can feel for yourself. The coccyx is the base of the spine and the ischial tuberosities are the parts of the bony pelvis that take the weight when you sit down. The gluteus maximus muscles (C), the levator ani muscle (D) and the anus (E) are the structures of the posterior perineum. A line between the ischial tuberosities passes just in front of the anus and marks the posterior boundary of the **uro-genital perineum**. The corpus spongiosum (F), containing the urethra, is covered laterally and anteriorly by the two corpora cavernosa (G), which are attached to the ischial tuberosities by the **ischio-cavernosus muscle** on each side. This attachment of the corpus cavernosum is called the **crus** (H).

Fig. 142 also shows the **dorsal vein** of the penis (I). The three corpora are arranged to form a cylindrical organ, the penis, which is about 6 inches long. They consist of erectile tissue which, when engorged with blood, stiffens and enlarges the penis, making it capable of entering the vagina. The corpus spongiosum ends as an enlargement called the glans penis; in it is the opening of the urethra, the **external urinary meatus**. The skin at the end of the penis is in the form of a fold, the **prepuce**, which covers the glans and is removed in **circumcision**.

Female Perineum

Fig. 143 shows the inferior aspect of the female perineum. The posterior perineum is similar to that shown in Fig. 142. The **perineal body** (A) lies on the posterior border of the uro-genital perineum and is the point of attachment for the **transverse perineal muscles** (B) and the levator ani, which, in the female, are important supports of the vagina (C). The vagina and the urethra (D) both open into the **vestibule** which is bounded on each side by the bulbo-spongiosus (E), erectile tissue similar to that of the male. At the posterior aspect of the bulbs are found a pair of glands, **Bartholin's glands** (F), which open into the

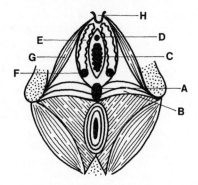

Fig. 143. Female perineum

vestibule and provide lubrication. Arising from the ischial rami are the two crura cavernosa (G) of the **clitoris** (H). The **glans clitoris** (H) lies just anterior to the opening of the urethra.

The skin covering the bulbs of the vestibule forms two thin delicate folds called the **labia minora.** They are enclosed by larger, hairy folds, the **labia majora,** which extend from the pubis anteriorly to meet posteriorly just over the perineal body. They are equivalent to the male scrotum. The vaginal opening is guarded in the virgin by a thin fold of mucosa, called the **hymen.** It has a perforation to allow the menstrual flow, and is usually ruptured during intercourse; after childbirth, only a few tags remain.

The Rectum and Anal Canal

The lower end of the intestinal tract is shown in Fig. 144. The **rectum** (A) is about 5 inches long and curves in front of the sacrum.

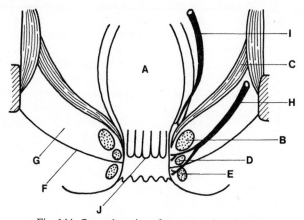

Fig. 144. Coronal section of rectum and anal canal

It ends in the **anal canal**, which is about $1\frac{1}{2}$ inches long and opens at the **anal orifice**. The rectum is lined by mucous membrane and is surrounded by two layers of muscle fibres: the inner circular layer and the outer longitudinal layer. The inner layer forms a powerful collar of involuntary muscle fibres, the **internal sphincter**, at the upper end of the anal canal. The **external sphincter** is composed of voluntary muscle and is arranged at three levels: the **deep sphincter** (B) blends with the levator ani (C) and the internal sphincter; the **superficial sphincter** (D) is attached to the perineal body anteriorly and the coccyx posteriorly; the **subcutaneous sphincter** (E) lies inferior to the perineal membrane (F). The **ischio-rectal fossa** (G) lies between the levator ani and the perineal membrane, and is filled with fatty tissue.

The lower half of the anal canal is lined by squamous epithelium which is continuous with that of the skin. The upper half is lined by mucous membrane which is arranged in vertical columns connected at their lower tips by flaps of mucous membrane, the **valves of Ball** (J). The line of valves is called the **pectinate line**; it marks the junction between the upper and the lower halves of the anal canal.

The upper half of the anal canal derives its blood supply from the **superior haemorrhoidal vessels** (I), which are branches of the superior rectal vessels. The veins drain into the **portal** system. The lower half of the anal canal derives its blood supply from the **inferior haemorrhoidal** vessels, which are branches of the pudendal vessels (H). These veins are part of the **systemic** system. The two systems anastomose in the walls of the anal canal. Portal hypertension may produce varicosities of the **haemorrhoidal veins**. These are called **haemorrhoids**, or piles. They can be produced in other ways, e.g. they are common in pregnancy because the foetal head exerts pressure on the veins draining the pelvis.

The lymphatic vessels above the pectinate line drain into the pelvis and abdomen; those in the lower half drain into the inguinal lymph nodes. The nerve supply above the pectinate line is from the autonomic nervous system; thus the mucous membrane can be injected, cut, or burnt without pain. The squamous epithelium of the lower half is supplied by somatic nerves, so it is sensitive to pain.

A hard faecal mass can tear the anal mucosa, and this is known as **fissure-in-ano**. A tear may lead to the spread of infection from the anal wall to the structures surrounding the anal canal. An **ischio-rectal abscess** is an abscess in the ischio-rectal fossa. If a connection remains open between the tear in the anal wall and the abscess, it is called a **fistula**.

CHAPTER EIGHT

THE HEAD AND NECK

Bones of the Skull

The individual bones of the skull can be distinguished more easily in an infant's skull (Fig. 145) than in an adult's. In this lateral view, the **frontal** (A), the **parietal** (B) and **occipital** (C) are shown forming the **vault** of the skull. The **temporal** bone consists of four parts: the thin,

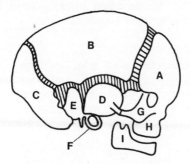

Fig. 145. Infant's skull (lateral)

flat **squamous** (D); the **mastoid** (E); the **tympanic** (F), which contains the opening for the ear; and the petrous, which cannot be seen from this aspect but which shows a pointed **styloid process** just posterior to the tympanic bone. The **zygomatic bone** (G) forms the prominence of the cheek, joining the squamous part of the temporal bone to the **maxillary bone** (H) which forms the upper jaw. the **mandible** (lower jaw) is shown at I.

Sutures

Many of the skull bones are paired, as seen in Fig. 146. The two frontal bones (A) are separated by membrane in foetal life, but they unite in about the fifth year and the junction usually becomes invisible. The paired parietal bones (B) are also separated by a membrane (D) at first; when they unite, the irregular line of the junction is called a **suture**. The membrane (E) between the frontal and parietal bones marks the line of the **coronal suture** which joins these bones in adult

131

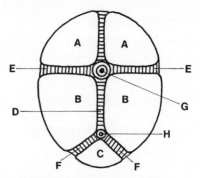

Fig. 146. Infant's skull (superior)

life. (Remember the names given to the anatomical planes, page 1.) The membrane (F) between the parietal bones and occipital bone (C) indicates the line of the **lambdoid suture,** named because of its resemblance to the Greek letter (lambda).

Fontanelles

In the foetus and infant, two relatively large areas of membrane are present in the vault of the skull. The **anterior fontanelle** (G in Fig. 146) can be felt easily, and pulsations corresponding to the heart beat can be seen as well as felt. This opening between the bones has usually closed by the age of two years. The smaller **posterior fontanelle** (H) closes shortly after birth.

Membrane Bones

Most of the bones forming the floor of the skull are formed in cartilage, but the bones forming the vault of the skull are formed in membrane. This type of bone, shown in Fig. 147, consists of two layers of hard compact bone: an outer layer (A) and an inner layer (B). The

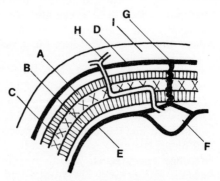

Fig. 147. Coronal section of skull

space between them is filled with cancellous bone (C) and contains red marrow throughout life. Adhering closely to the outer layer is periosteum (D) which continues through the suture line (G) to join the fibrous membrane (E), called the **dura mater**, lining the brain and spinal cord. In certain areas the dura mater splits to enclose a **venous sinus** (F). These sinuses drain most of the blood from the brain. Veins (H) in the scalp communicate with the venous plexus in the cancellous bone which empties into the venous sinuses. Infection in the scalp can therefore spread to the bones (**osteomyelitis**) and even the cranial cavity (**meningitis**).

The Adult Skull

In the adult skull (Fig. 148) the membrane between the individual bones has been replaced by bone, and only the suture lines can be

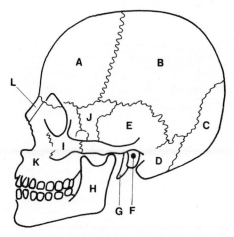

Fig. 148. Adult skull (lateral)

seen. The coronal suture separates the frontal bone (A) from the parietal bone (B). One limb of the lambdoid suture separates the parietal bone from the occipital bone (C). Four parts of the temporal bone can be seen: the squamous (E); the mastoid (D), with its process which can be felt just behind the ear; the tympanic (F), which forms most of the bony external auditory canal; and the styloid process (G) of the petrous temporal. The bone labelled J is called the **greater wing of the sphenoid**. The right and left greater wings are attached to the body of the sphenoid bone which lies in the floor of the skull.

Bones of the Face

The zygomatic bone (I) is attached to the temporal bone posteriorly, and anteriorly to the maxilla (K) and frontal bones. The short **nasal**

bone (L) is paired. The mandible (H) pivots against the inferior surface of the temporal bone during the movements of the mouth.

The mandible, as seen from the lateral aspect (Fig. 149*a*), consists of a **ramus** (A) and a **body** (B). The upper end of the ramus has a **condyloid process** (C) consisting of a head, which forms part of the joint, and a neck. The **coronoid process** (D) receives the insertion of the large temporalis muscle.

The angle of the mandible (E) is formed where the posterior border of the ramus meets the inferior border of the body. The bony opening in the body of the mandible (F) is an exit for the terminal branch of the **inferior alveolar nerve**, which runs within the mandible to supply the

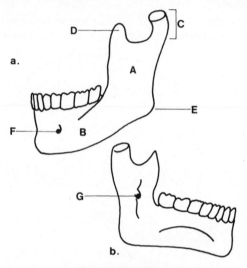

Fig. 149. Mandible

lower teeth. The bone of the mandible is hard and dense, so a local anaesthetic placed in the gum has little chance of reaching the nerve supplying that tooth. Thus the nerve itself must be anaesthetized before it enters the mandible. The bone of the maxilla is much less dense (in fact it is honeycombed with small apertures), so an upper tooth can be easily anaesthetized by infiltrating the local anaesthetic into the gum near the root of the tooth.

On the medial surface of the mandible (Fig. 149*b*), the opening in the bone (G) is the entrance for the inferior alveolar nerve, and this is the usual site for the injection of local anaesthetic. The upper border of the body of the mandible has 16 dental sockets, or **alveoli**. There are 32 permanent teeth, made up in each half jaw of: two **incisors**, one **canine**, two **premolars**, and three **molars**. The third molar, called the wisdom tooth, does not appear until early adult life.

Cranial Cavity: The Fossae

The brain fits into the skull exactly, and the floor of the cranial cavity consists of three **fossae**, or depressions, which contain three parts of the brain (Fig. 150). The **anterior cranial fossa** (A) is formed by the bony roof of the orbit and the nasal cavity; the frontal lobe of

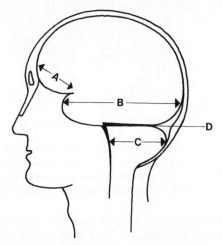

Fig. 150. Sagittal section of cranial cavity

the brain rests on it. The **middle cranial fossa** (B) extends to the posterior surface of the skull. The anterior part of the floor is bony, and the rest is formed by a tough fibrous membrane (D). The temporal lobe of the brain rests on the anterior part of the cavity, while the occipital lobe rests on the membrane, which is called the **tentorium cerebelli.** The **posterior cranial fossa** (C) lies inferior to the tentorium and contains the cerebellum and the hind brain.

When the vault (roof) of the skull is removed, along with the brain, the bones forming the floor of the cranial cavity are revealed as in Fig. 151. Anteriorly, the floor of the anterior cranial fossa is formed by the **orbital plate** of the frontal bone (A). In the midline, the **cribriform plate** of the **ethmoid bone** (B) is perforated to allow the olfactory nerves to pass from the cranial cavity to the nasal cavity where they respond to odours. Leakage of cerebrospinal fluid into the nasal cavity may result from a skull fracture.

The Sphenoid

The sphenoid bone (Fig. 151) consists of the **lesser wings** (C) separated from the greater wings by the **superior orbital fissure** (J). The **pituitary fossa** (D) on the body of the sphenoid is a small hollow holding the pituitary gland, which is attached by a stalk to the inferior

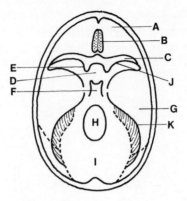

Fig. 151. Floor of cranial cavity

surface of the brain. The fossa is surrounded by four bony prongs. The anterior prongs are part of the lesser wings and are called the **anterior clinoid processes** (E). The **posterior clinoid processes** (F) are part of the **dorsum sellae** (F), which is the raised posterior part of the body of the sphenoid.

The **petrous** parts of the temporal bone (G) are visible posterior to the greater wings of the sphenoid. This bone is marked by a ridge (K) which gives attachment to the tentorium cerebelli. The part anterior to this ridge forms part of the middle cranial fossa, and the other part forms the side wall of the posterior cranial fossa. The **foramen magnum** (H) in the occipital bone (I) is where the spinal cord, the continuation of the brain stem, leaves the cranial cavity and enters the spinal column.

The sphenoid bone is difficult to visualize from a diagram of the skull. Fig. 152 shows the posterior aspect of the bone, separated from the rest of the skull. The **body** (A) contains a large **air sinus** (B) and a depression in its roof is the pituitary fossa (C). The lesser wing (E) is separated from the greater wing (D) by a gap, the superior orbital fissure (F). Except for the optic nerve, which has its own canal, most of

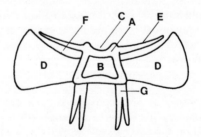

Fig. 152. Sphenoid

the vessels and nerves going to the orbit (eye-socket) pass through this fissure. From the inferior surface of the body two processes (G), called the **pterygoid plates**, project downwards into the nasopharynx.

The Vertebral Column

The vertebral column (spinal column) is made up of 33 vertebrae which articulate together to form a rigid, flexible support for the spinal cord. The seven **cervical vertebrae** are found in the neck; these are represented as C1, C2, ... from above downwards. Fig. 153 shows the top three cervical vertebrae. The upper surface of the **atlas**

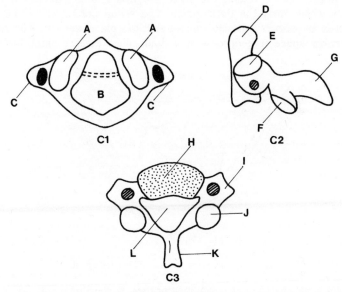

Fig. 153. Vertebrae

(C1) has two articular facets (A) which are part of the joint with the occipital bone. The spinal cord passes through the **vertebral canal (B)**. The groove on the **transverse process** (C) is for the vertebral artery which passes upwards into the skull through the foramen magnum.

A side view of the **axis** (C2) shows the **odontoid process** (D) which projects upwards and fits into the atlas. The upper facets of the axis (E) articulate with the atlas, and the lower facets (F) with C3. Nodding of the head and a small degree of lateral flexion take place at the joint between the skull and the atlas. Rotation of the head occurs around the odontoid process. Flexion and extension of the neck involve sliding movements between all the cervical vertebrae. The axis has a **spine** (G) on its posterior aspect, as do all the cervical vertebrae except the atlas.

The upper aspect of C3 shows the body (H), the spine (K), the vertebral canal (L) and the lateral processes (I) which are pierced for the passage of the vertebral artery. The other cervical vertebrae have the same basic pattern.

Membranes of the Cord and Brain

The brain and spinal cord are invested in three layers of membrane, called **meninges.** A cross section of the spinal cord in the vertebral canal (Fig. 154) shows the arrangement of the meninges. The **dura mater** (A) is a tough fibrous sheath which lines the vertebral canal. The dotted line represents the **arachnoid mater,** a thin delicate membrane which lines the dura mater. The spinal cord itself is closely covered by another delicate membrane, the **pia mater** (B). The **subarachnoid space** (C), filled with a fluid called **cerebrospinal fluid** or

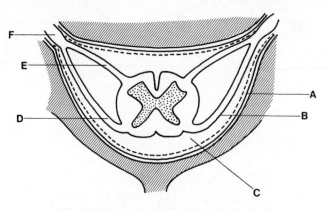

Fig. 154. Transverse section of vertebral canal

C.S.F., separates the two inner membranes. A specimen of C.S.F. can be obtained by inserting a hollow needle between two lumbar vertebrae into the sub-arachnoid space. This procedure, **lumbar puncture,** is necessary in certain diseases such as meningitis (inflammation of the meninges).

The anterior or ventral root (E) carries motor nerves from the spinal cord and links up with the posterior root (D) to form the spinal nerve (F). The spinal nerve leaves the vertebral column between the transverse processes and is still invested by the membranes.

The coverings of the spinal cord are continuous with the coverings of the brain. The dura mater which lines the cranial cavity forms two layers in some places: the outer layer adheres to the skull, and the inner layer splits away to form venous sinuses and supporting structures, as shown in Fig. 155. The **falx cerebri** (C) dips down between the two cerebral hemispheres, and the **tentorium cerebelli** (D) separates the

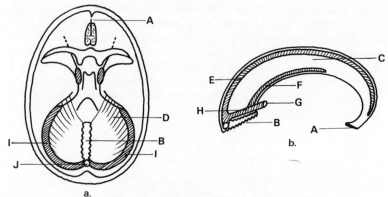

Fig. 155. Attachment of dura mater

cerebellum from the cerebrum. The tentorium cerebelli is shown in position lying across the posterior cranial fossa. The falx cerebri has been removed. To place it in position, superimpose the areas A and B on the corresponding areas in the cranial cavity, so that the falx is at right-angles to the tentorium. (Trace it and cut the tracing out to get a three-dimensional impression.)

The falx contains a large **superior sagittal sinus** (E) and a smaller **inferior sagittal sinus** (F). A large vein from the brain (G) joins the inferior sagittal sinus to form the **straight sinus** (H). Blood from these sinuses now enters the **right** and **left transverse sinuses** (I) at J. These sinuses lie in the outer edge of the tentorium cerebelli, which has a notch in its anterior edge for the midbrain.

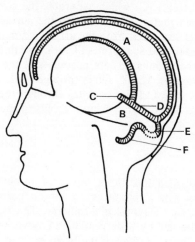

Fig. 156. Venous sinuses

As seen in their normal relationship from the lateral aspect (Fig. 156), the inferior sagittal sinus in the lower border of the falx (A) joins the great cerebral vein (C) to form the straight sinus (D) which lies at the junction of the falx with the tentorium (B). The superior sagittal sinus, in the upper margin of the falx, joins the straight sinus to empty into the transverse sinus (E). This runs horizontally and then dips into the posterior cranial fossa, where it is called the **sigmoid sinus** (F), to empty into the internal jugular vein.

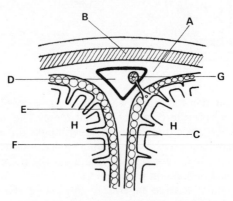

Fig. 157. Coronal section of falx cerebri

Fig. 157 is a coronal section which shows the arrangement of the meninges at the falx cerebri. The dura mater (A) lines the vault of the skull (B) and splits to enclose the superior sagittal sinus (D). It reunites to form the falx cerebri (C) which separates the two cerebral hemispheres (H). The pia mater (F) adheres to the surface of the cerebral hemispheres and is separated from the arachnoid mater (E) by a relatively large sub-arachnoid space full of C.S.F. and fine weblike strands. The **arachnoid granulations** (G) are processes of the arachnoid mater into the venous sinuses. C.S.F. filters through these granulations into the venous blood. The C.S.F. circulates through the central nervous system and carries waste matter from the cells of the brain and spinal cord to the blood stream.

THE BRAIN

The brain is divided into three parts: the **forebrain**, the **midbrain** and the **hindbrain**. The forebrain is the largest part and most of it is composed of the two **cerebral hemispheres**. The right hemisphere is separated from the left by a deep midline groove, which contains the falx cerebri. The lateral surface of the brain is shown in Fig. 158, from the left.

Cerebral Hemispheres

The cerebral hemispheres possess an outer layer of grey matter, composed of nerve cells, which is known as the **cortex**. The cerebral cortex covers the large mass of white matter, composed of myelinated nerve fibres, which connect the hemispheres with other parts of the central nervous system. The surface of the cerebral hemispheres is folded into wrinkles, or convolutions; the ridges are called **gyri** (singular, *gyrus*), and the fissures which separate them are called **sulci** (singular, *sulcus*). These convolutions increase the surface area, or cortex, without a great increase in the volume of the brain. The sulci provide landmarks for the division of the cerebral hemispheres into **lobes**, shown in Fig. 158.

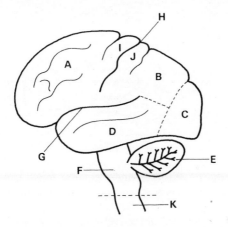

Fig. 158. Brain (lateral)

The central sulcus (H) separates the **frontal lobe** (A) from the **parietal lobe** (B). The lateral sulcus (G) lies above the **temporal lobe** (D), and the **occipital lobe** (C) is separated from the parietal lobe by a fissure on the medial aspect of the hemisphere.

The gyrus immediately in front of the central sulcus is called the **pre-central gyrus** (I). This area is known as the **motor cortex**, as it is composed of nerve cells which initiate voluntary movement in the opposite half of the body. The nerve fibres from the left motor cortex cross to the right side of the central nervous system and supply muscles on the right side of the body. This crossing, or **decussation**, of the nerve fibres means that an injury, such as haemorrhage, to the left side of the brain may result in paralysis on the right side of the body.

The **post-central gyrus** (J) is concerned with the recognition of sensations, such as touch, pain and temperature. The sensory fibres also cross over in the central nervous system so that sensations from the left side of the body are relayed to the right post-central gyrus, and vice versa. Each of the special senses is relayed to a specific area of the cortex. The **auditory pathway** from the ear finishes in the part of the temporal lobe which lies just below the lateral sulcus. The **visual pathway** from the eye ends in the occipital lobe.

The midbrain is hidden by the cerebral hemisphere in the lateral view. The hindbrain consists of the **pons** and **medulla** (F) and the **cerebellum** (E). The spinal cord (K) is continuous with the medulla and begins at the foramen magnum, which is indicated by the dotted line.

The medial surface of the right cerebral hemisphere is shown in Fig. 159. The four lobes are indicated and the **parieto-occipital sulcus** (E)

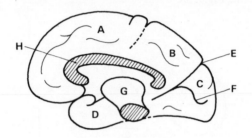

Fig. 159. Right cerebral hemisphere (medial)

which separates the parietal and occipital lobes can be seen. The **calcarine sulcus** (F) is very deep and its sides contain the visual area of the occipital lobe.

A large mass of nerve fibres (H) connects the two cerebral hemispheres and is called the **corpus callosum**. Parts of the forebrain called the **thalamus** and **hypothalamus** are situated below the corpus callosum and above the midbrain (G), which forms a bridge between the forebrain and the rest of the central nervous system.

Parts of the Brain

In Fig. 160 a sagittal cut has been made through the brain, starting in the deep midline groove between the hemispheres and continuing through the corpus callosum (K) and the remaining parts of the brain.

Starting at the spinal cord (D) and moving upwards is the **medulla oblongata** (C). The dotted line in Fig. 160 indicates the position of the foramen magnum, which is the dividing line between the spinal cord and the hindbrain. The cerebellum (E) lies posterior to the pons (B). The midbrain (A) connects the pons with the forebrain. The **pituitary gland** (F) is connected by a stalk to the hypothalamus; it is part of the endocrine system of glands which produce hormones and release them

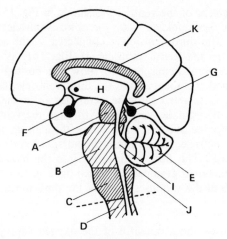

Fig. 160. Sagittal section of brain

into the blood stream. The **pineal gland** (G) is connected by a stalk to the thalamus. The function of the pineal gland is unknown.

Ventricles

The brain starts in the embryo as a tube of nerve tissue. As it develops and gets more complicated, parts of the tube dilate and form **ventricles** which are full of cerebrospinal fluid (C.S.F.). The **third ventricle** (H) is in the forebrain where it separates the right and left thalami. It communicates with the **fourth ventricle** (I) in the hindbrain by a narrow channel known as the **aqueduct of the midbrain**. The fourth ventricle communicates with the central canal of the spinal cord. Each cerebral hemisphere contains a ventricle which is known as the **lateral ventricle**. These communicate with the third ventricle.

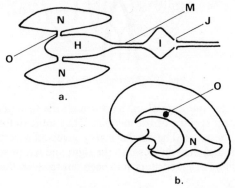

Fig. 161. Ventricles

The arrangement of the ventricles is shown in Fig. 161*a*. The lateral ventricles (N) communicate with the third ventricle (H) by small openings (O). The third ventricle is connected by the aqueduct of the midbrain (M) to the fourth ventricle (I). Fig. 161*b* is a medial view of the right cerebral hemisphere to show the position of the lateral ventricle (N) and the opening into the third ventricle (O).

The C.S.F. is secreted into the ventricles from the blood vessels in their walls. It circulates very slowly through the ventricular system and passes into the sub-arachnoid space through small openings (J) in the fourth ventricle. (The sub-arachnoid space surrounds the brain and spinal cord.) The fluid is absorbed into the blood stream again through the arachnoid granulations which have already been described as projecting into the venous sinuses of the dura mater.

Blood Supply

The brain uses about one-fifth of the oxygen supply of the body, and nerve cells die if they are deprived of oxygen for longer than a few seconds. The blood supply to the brain is thus essential for life and it is very large, about one-sixth of the cardiac output. The major arteries supplying the brain are shown in Fig. 162. The two **vertebral**

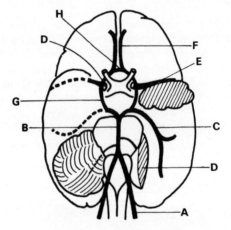

Fig. 162. Arteries of the brain

arteries (A) enter the cranial cavity through the foramen magnum to lie on the ventral surface of the medulla. They unite to form the **basilar artery** (B) on the pons. The basilar artery gives off several branches to the cerebellum and then divides into the **right** and **left posterior cerebral arteries** (C). The left cerebellum has been cut to show the left posterior cerebral artery.

The other major artery to the brain is the **internal carotid,** the cut end of which can be seen in Fig. 162 at D. This artery divides into a **middle cerebral artery** (E) and an **anterior cerebral artery** (F). A small branch from the internal carotid artery, the **posterior communicating artery** (G), joins with the posterior cerebral branch of the basilar artery. The anterior cerebral arteries are joined by a short **anterior communicating branch** (H). Observe how the arteries thus form a circle which is known as the **circle of Willis.** Begin at the basilar (B), then pass to the posterior cerebral (C), to the posterior communicating (G), to the internal carotid (D), to the anterior cerebral (F), and across the anterior communicating branch (H) to the opposite side of the circle. These arteries lie in the sub-arachnoid space. They sometimes develop **aneurysms,** a dilation of a segment of the artery. The aneurysm may press on adjacent nerves and damage them, or may rupture and produce a **sub-arachnoid haemorrhage,** which may be fatal.

The branches of the middle cerebral artery supply the motor and sensory tracts of the forebrain. If they become blocked by a blood clot (thrombosis) or rupture and produce haemorrhage in this area, the result is a combination of paralysis and sensory loss on the opposite side of the body, which is called a 'stroke'.

Cranial Nerves

The nerves leaving the brain, and the blood-vessels supplying the brain, pass through openings called foramina (singular, foramen) in the base of the skull (Fig. 163). The first cranial nerve is the **olfactory**

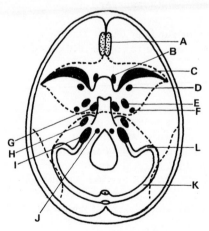

Fig. 163. Foramina of the base of the skull

nerve which arises in the nose and passes to the **olfactory bulbs** on the cribriform plate (A) of the ethmoid bone. The second cranial nerve is the **optic nerve** which arises in the retina of the eye and passes through the **optic canal** (B) to reach the brain. The third (**oculomotor**), fourth (**trochlear**) and sixth (**abducens**) cranial nerves are all motor nerves arising in the brain and supplying the muscles of the eye. They reach the orbit (eye-socket) by passing through the superior orbital fissure (C) which also contains some of the blood vessels supplying the orbit.

The fifth cranial nerve, **trigeminal**, is largely a sensory nerve supplying the skin of the face and some underlying structures such as eyelids, nasal cavity, lips and teeth. The ganglion of the trigeminal nerve corresponds to the sensory or posterior ganglion of a spinal nerve. It is called the **semi-lunar ganglion** and lies on the petrous part of the temporal bone. The sensory fibres which join to form the ganglion consist of three distinct branches: ophthalmic, maxillary and mandibular. The **ophthalmic** division passes through the superior orbital fissure (C) to reach the orbit and the overlying structures of the orbit. The **maxillary** division passes through the **foramen rotundum** (D) to supply the structures in the maxillary bone and the skin covering it. The **mandibular** division passes through the **foramen ovale** (E) to mandibular structures, including teeth, tongue, lower lip and skin. The mandibular division also carries motor fibres supplying the muscles of mastication, with the exception of the buccinator muscle which is supplied by the facial nerve.

The seventh (**facial**) and eighth (**auditory**) nerves leave the cranial cavity together via the **internal auditory meatus** (H). The ninth (**glossopharyngeal**), tenth (**vagus**) and eleventh (**accessory**) nerves leave through the **jugular foramen** (I). The twelfth (**hypoglossal**) nerve leaves through the **anterior condylar canal** (J). The functions of these nerves will be discussed later with reference to the structures they supply.

Blood-vessels

The **foramen spinosum** (F) transmits the **middle meningeal artery** which supplies the dura mater and the skull bones. The internal carotid artery enters the cranial cavity through the **carotid canal** (G) which runs through the petrous part of the temporal bone close to the ear. The blood from the brain is drained into the venous sinuses. The transverse sinuses (K) are shown in Fig. 163 turning medially to form the sigmoid sinuses (L). These sinuses empty into the beginnings of the **internal jugular veins** which leave through the jugular foramina (I).

Attachments

The attachments of the cranial nerves to the ventral surface of the brain are shown in Fig. 164 and are summarized in Table 1. The cranial nerves are usually given Roman numerals.

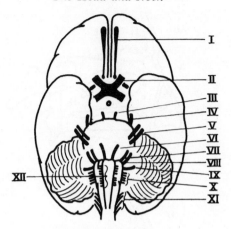

Fig. 164. Cranial nerves

Table 1. Attachments of the Cranial Nerves

	Nerve	Attachment
I	Olfactory	Cerebrum
II	Optic	Cerebrum
III	Oculomotor	Midbrain
IV	Trochlear	Midbrain
V	Trigeminal	Pons
VI	Abducens	Junction of pons and medulla
VII	Facial	Junction of pons and medulla
VIII	Auditory	Junction of pons and medulla
IX	Glossopharyngeal	Medulla
X	Vagus	Medulla
XI	Accessory	Medulla (and branches from spinal cord)
XII	Hypoglossal	Medulla

THE NOSE AND SINUSES

The nose is a passageway divided into right and left **nasal cavities** by a median **septum**. Air enters through the nostril and traverses the nasal cavity to enter the pharynx. The bones which make up the nose are shown in Fig. 165. The side wall of the right nasal cavity (Fig. 165a) consists of several bones and pieces of cartilage. The external nose is formed by paired nasal bones (A) which are attached to the frontal bone (B). This junction is known as the root of the nose. The inferior half of the external nose is formed by cartilages (C) that allow some mobility.

The ethmoid bone (D) forms the upper part of the wall. This is a thin bone containing air sinuses and a cribriform plate (E) which is perforated to allow the passage of olfactory nerves to the cranial cavity. The

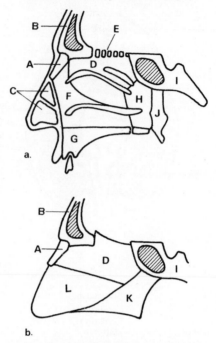

Fig. 165. Nasal cavity (lateral) and septum

lower part of the wall is formed by the maxillary bone (F), which also forms the anterior part of the **hard palate** (G) separating the floor of the nasal cavity from the mouth. The posterior part of the hard palate is formed by the **palatine bone** (H) which extends upwards to join the sphenoid bone (I) in the base of the skull. The **medial pterygoid plate** (J) projecting down from the sphenoid completes the lateral wall of the nasal cavity.

The frontal bone and the sphenoid bone both contain large air sinuses which communicate with the nasal cavity. Projecting into the nasal cavity from the lateral wall are three shelf-like structures called the **turbinates** that increase the surface area. When air is breathed in through the nose, it is made warm and damp by its passage over the respiratory surface of the turbinates before it enters the trachea. The lower and middle turbinates are larger than the superior.

Septum

The dividing wall or septum between the two nasal cavities is shown in Fig. 165 b. The nasal, frontal, ethmoid and sphenoid bones are all involved in the upper part of the septum. The **vomer bone** (K) forms the posterior part of the septum; cartilage (L) anterior to the vomer and to the vertical plate of the ethmoid (D) completes the septum. Deviation of the septum is a common cause of nasal obstruction. Most cases are the result of injury, but sometimes deviation slowly develops during childhood as the constituent parts grow at different rates.

Accessory Nasal Sinuses

These sinuses are air-containing sacs in the frontal, ethmoidal, sphenoid and maxillary bones. They are lined by mucous membrane, continuous with that of the nasal cavities, and their openings into the nasal cavities are narrow and easily blocked. The mucus secreted by the membrane of the nose and sinuses is normally moved through the nasal cavity towards the pharynx, and then swallowed. The production of mucus is greatly increased during infection, or smoking, and it may drip from the nostrils unless removed by the use of a handkerchief.

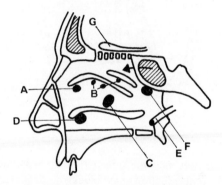

Fig. 166. Openings into the nasal cavity

The openings in the lateral wall of the right nasal cavity are shown in Fig. 166. The **frontal sinus** drains into the nose under the middle turbinate (A). The **ethmoidal sinuses** are a honeycomb of air cells in the medial walls of the orbits. They drain below the superior and middle turbinates (B). The **sphenoidal sinus** is in the body of the sphenoid bone and it drains through an opening indicated by the arrow. The largest sinus related to the nose is in the **maxillary** bone and is triangular in shape. It drains just above the inferior turbinate (C). This opening is high on the medial wall of the sinus, so gravity does not assist in drainage.

The **lacrimal sac** in the orbit contains the fluid that bathes the eyes; it drains into the **naso-lacrimal duct** which opens below the inferior turbinate (D). The middle ear is connected with the naso-pharynx by the **Eustachian tube** (E) which opens in line with the inferior turbinate. An opening in the upper part of the palatine bone, just below the sphenoid, allows the entry of the major nerves and blood-vessels of the nasal cavity. This opening is the **spheno-palatine foramen** (F). The olfactory bulb (G) lies just above the cribriform plate.

The relationship of the nasal cavities and sinuses is shown in Fig. 167, which is a coronal section through the nose. The nasal cavities (A) are separated by the septum (B). Projecting from the lateral wall of each cavity are the turbinates: the superior turbinate (D), the middle turbinate (E) and the inferior turbinate (F). The space just below a turbinate is called a **meatus** (C). The maxillary sinus (G) drains into the

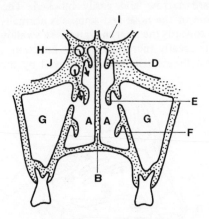

Fig. 167. Coronal section of the nasal sinuses

middle meatus. It may become infected, either from the nasal cavity or from a tooth, and infection may block the drainage of the sinus; the result is pain in the cheek, and an opening may have to be made surgically into the inferior meatus.

The roof of the maxillary sinus is the floor of the orbit (J), and the ethmoidal sinuses (H) are medial to the orbit. Infection in the sinuses may cause pain and swelling between the eyes. The frontal lobes of the brain (I) lie just above the ethmoid bone. Infection in the sinuses may spread to the cranial cavity (**cerebral abscess**), and fracture of the ethmoid bone may result in the leakage of C.S.F. down the nose. The frontal sinuses are anterior to the ethmoid and lie above the orbits. They are also closely related to the frontal lobes of the brain.

Blood Supply

There are four main arteries to the walls of the nasal cavity. They anastomose on the anterior part of the septum, which is frequently the site of **epistaxis** (nose-bleeds). The arteries are the **anterior ethmoidal**, a branch of the ophthalmic; the **spheno-palatine**, which branches through the spheno-palatine foramen from the maxillary; the **greater**

palatine, which passes upwards through the palate; and branches from the artery which supplies the upper lip. The arteries run horizontally in the mucous membrane covering the turbinates, where they warm up the inspired air as it passes over them. Infection in the nasal cavity (**coryza**) causes swelling of the arteries and the mucous membrane, which blocks the air-passage. When substances containing adrenalin are applied to the mucous membrane the blood vessels are constricted, but prolonged use of such nasal decongestants may damage the nasal mucosa.

Epistaxis from the anterior part of the nasal septum can usually be controlled by pressure or by plugging the nasal cavity with gauze soaked in a mixture containing adrenalin. Epistaxis from the posterior part of the nasal cavity may be due to hypertension and may be more difficult to control. It may be serious enough to require treatment in hospital, including blood transfusion.

Coryza (the common cold) may result in infection spreading to the sinuses which are connected with the nasal cavity. The infection may also spread up the naso-lacrimal duct to the eye, which becomes red and sticky. The Eustachian tube may carry the infection to the middle ear, which may result in temporary deafness, or earache, and odd sounds in the head.

THE EYE

The bony socket of the eyeball, called the **orbit**, is cone-shaped. In the horizontal section (Fig. 168*a*), looking into the right orbit from above, the medial wall (A) runs forward and the lateral wall (B) makes an angle of 45° with it. Because of this inclination, the lateral surface of the eyeball is exposed and more accessible for surgery. The optic nerve, carrying sensation from the retina, passes through the apex of the orbit on its way to the brain. The **eyeball**, about one inch in diameter, projects forward from the orbit; the space behind it is filled with muscle, nerves and fat. The sunken eyes of a starving person result from a decrease in orbital fat. The sunken eyes of an infant with gastro-enteritis result from a decrease in fluid in the contents of the orbit.

Muscles

Movements of the eye result from the action of six muscles in the orbit. The **medial rectus muscle** and the **lateral rectus muscle** (Fig. 168*a*) arise from the apex of the orbit and are inserted into the respective surfaces of the eyeball. The other muscles are shown in Fig. 168*b*, which is a sagittal section through the orbit. The **superior rectus** (C) and the **inferior rectus** (E) are inserted on the upper and lower surfaces of the eyeball. The **superior oblique muscle** (F) runs through a sling or

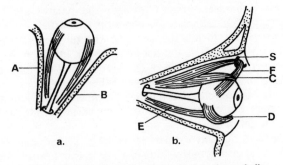

Fig. 168. Orbit and muscles attached to the eyeball

pulley (S) made of fascia, and bends back to be inserted obliquely on the upper surface of the eyeball. The inferior oblique muscle (D) inserts on the lateral surface of the eyeball.

The oculomotor, or third cranial, nerve supplies the muscles of the eyeball except the superior oblique and the lateral rectus. The superior oblique muscle is supplied by the trochlear, or fourth cranial, nerve; the lateral rectus is supplied by the abducens, or sixth cranial, nerve.

Movements

Fig. 169 shows the movements of the right eye. The superior rectus (SR) and inferior oblique (IO) muscles elevate the eyeball (Fig. 169a). The superior oblique (SO) and inferior rectus (IR) muscles depress the eyeball (Fig. 169b). The superior rectus (SR) and superior oblique (SO) muscles produce medial rotation of the eyeball (Fig. 169c). The

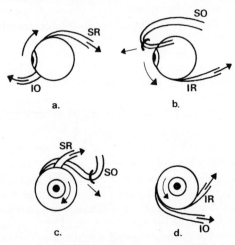

Fig. 169. Movements of the eyeball

opposite movement, lateral rotation, is produced by the inferior rectus (IR) and the inferior oblique (IO) muscles (Fig. 169*d*).

The medial rectus muscle acts alone to produce adduction, i.e. turning the eyeball towards the nose. The lateral rectus acts alone to produce the opposite movement, abduction, i.e. turning the eyeball from the nose.

Normally the muscles of both eyeballs act together so that both eyes face in the same direction. Weakness of one of the eye muscles results in a deviation of one eye, which is a **squint**. This may be corrected by operating to shorten the weak muscle or muscles.

Layers of the Eyeball

For simplicity, each of the three layers of the eyeball is shown separately in Fig. 170. The outer layer (Fig. 170*a*) is the **sclera** (A) and it forms the white of the eye. The anterior part of the sclera becomes clear and transparent to form the **cornea** (B). The **limbus** (D) is the circular junction between sclera and cornea. Within the sclera near the limbus, is a circular channel (C) called the **canal of Schlemm**. This encircles the eyeball and is concerned with drainage of fluid inside the eyeball. The posterior wall of the sclera has sieve-like openings (E) for the passage of the fibres of the optic nerve. A part of the sclera (F) continues as a sheath for the optic nerve and is continuous with the dura mater of the brain.

The middle coat, called the **choroid**, is shown in Fig. 170*b*. This is the vascular layer of the eyeball. A thickening (A) forms the **ciliary**

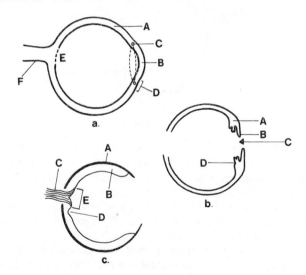

Fig. 170. Layers of the eyeball

body which contains the muscles necessary for accommodation, described later. The anterior part of the choroid layer forms a disc, the **iris** (B), with a hole in its centre, the **pupil** (C). Small extensions from the ciliary body, behind the iris, are called **ciliary processes** (D), and contain veins concerned with the production of aqueous fluid.

The lining of the eyeball (Fig. 170c) is the **retina**. It has an outer pigmented layer (A) and an inner layer of nerve cells and fibres (B). Detachment of the retina, sometimes due to injury, is the separation of the two layers. The nerve fibres converge to form the optic nerve (C) which leaves the eyeball at its posterior pole (E). This exit is called the **optic disc**; since it contains no nerve cells, it is the 'blind spot' of the retina. Near the lateral side of the optic disc is a small spot, called the **macula lutea**, which can be seen with the aid of an instrument for inspecting the inside of the eyeball (**ophthalmoscope**). In the centre of the macula lutea a thinning of the retina forms a small depression called the **central fovea** (D). The central fovea is the most sensitive area of the retina, and light rays are focused on to this spot when we look directly at an object.

Aqueous Humour

In Fig. 171 the three layers of the eyeball have been put together in a sagittal section. The sclera and cornea (A) enclose the choroid (B) and the retina (C). The **conjunctiva** (D) covers the cornea. The

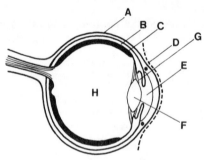

Fig. 171. Sagittal sections of eyeball

lens (F) is held in position by the suspensory ligaments arising from the ciliary body. The space (E) between the iris and cornea, filled with **aqueous humour,** is the **anterior chamber**. The **posterior chamber** (G) is the fluid-filled space between the lens and the iris. The ciliary processes in the posterior chamber produce the aqueous humour which flows through the pupil into the anterior chamber and is then absorbed into the canal of Schlemm near the angle between the iris and rim of the cornea. Excess production of this fluid, or inadequate drainage, raises the pressure within the eyeball. This condition, called **glaucoma,** must be treated rapidly (either by drugs or by surgery) or it will result in permanent blindness. The first symptom of increased pressure is a typical disturbance of vision.

The space H is filled with a clear gel, the **vitreous humour**, which helps to maintain the spherical shape of the eyeball.

Accommodation

In the relaxed eye, light rays from a distant object are brought to a focus on the retina. To accomplish this, the light rays are refracted (bent) by the cornea, aqueous humour, lens and vitreous humour. Most of the refraction takes place at the cornea; the main function of the lens is to alter the angle of refraction. When an object is very near the eye, the light rays from it would be brought to a focus behind the retina in the relaxed eye. However, an automatic mechanism within the eyeball increases the refractive power of the lens in this situation, so the light rays from the nearby object are brought to a point on the retina. This mechanism for near vision is called **accommodation.**

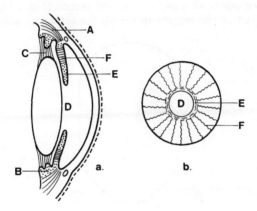

Fig. 172. Lens and iris

The lens in youth is normally rather soft and its shape is partly determined by the tensions exerted on it. Tension is exerted on the lens by the suspensory ligaments attached to its periphery (Fig. 172*a*). Contraction of the radiating fibres (A) pulls the ciliary body forwards; contraction of the circular fibres (B) brings the ciliary body closer to the edge of the lens, like the pull on a purse-string. The distance between the points of attachment of the suspensory ligaments (C) is shortened, and so the tension they exert on the lens is decreased. The lens consequently bulges forward more, and increases its convexity and hence its refractive power. Light rays from a near object are thus bent at a greater angle, to come to a point of focus on the retina.

Light Reflex

The muscles of accommodation receive their motor supply from parasympathetic nerve fibres travelling in the oculomotor nerve. Look into a mirror in a darkened room and you will notice that your pupils are quite large. Shine a torch into them and you will see the pupils contract. This constriction of the pupil is called the **light reflex**. The iris can change the size of its aperture, the pupil (D), by the action of two muscles. The circular fibres (E) at the periphery of the pupil contract, and so constrict the pupil. The radiating muscle fibres (F) contract to dilate the pupil. The iris is shown from the front in Fig. 172*b*. The circular fibres which constrict the pupil are supplied by parasympathetic nerves which are active in accommodation. The radial muscle fibres which dilate the pupil are supplied by sympathetic nerves. As usual, the actions of these two parts of the autonomic system are in opposite directions.

The Eyelids

The **orbicularis oculi** muscle closes the eyelids (Fig. 173). The more peripheral part of the muscle surrounds the orbit (A), and the central

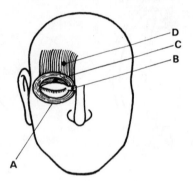

Fig. 173. Orbicularis oculi

part (C) is found within the eyelids. The muscle is arranged as an oval sweep of fibres which are attached to a fibrous band at the inner angle of the eye (B). The **frontalis muscle** (D) in the forehead helps to raise the upper lid as well as the eyebrow. These muscles are supplied by the facial nerve. If the facial nerve is paralysed, either following a stroke or as a result of Bell's palsy, the eyelids on the affected side cannot be closed.

The structures found within the eyelids are shown in Fig. 174. The outer surface of the lids is covered by skin (A) which is continuous with the conjunctiva (B) covering the inner surface. The conjunctiva is reflected on to the surface of the visible part of the eyeball. The space between lid and eyeball is called the **conjunctival sac** (C). Secretions from the **lacrimal gland** (D) enter this sac through several ducts and form the tears. This fluid is normally drained into the lacrimal sac (not shown), and from thence into the nose. It overflows from the eyes if produced in excess—as in misery, cold weather or exposure to acrid vapours —or if the naso-lacrimal duct is blocked, as in the common cold.

The rigidity of the lids is caused by plates of cartilage; the **superior tarsal plate** (E) is larger than its counterpart in the lower lid. Lying anterior to the tarsal plate, the cut ends of the orbicularis oculi muscle are shown (F). Posterior to the plate are the **tarsal glands** (G) which, together with the tears, provide lubrication. At the edge of the lids are found the eyelashes and the **ciliary glands**. A stye is an inflamed ciliary gland.

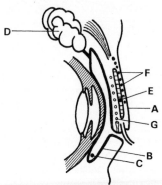

Fig. 174. Sagittal section of eyelids

When looking for a foreign body in the eye, place a matchstick against the upper lid just below the eyebrow and parallel with it. Using your other hand, grasp a few eyelashes and raise the lid. The bending and eversion of the lid take place above the upper edge of the tarsal plate, and the rigidity of the plate will keep the lid everted. Thus a good view is obtained of the upper conjunctival sac. The space behind the lower lid is seen very easily.

The chief muscle which raises the upper lid is found in the orbit just above the superior rectus muscle. It is the **levator palpebrae** and is supplied by both oculomotor and sympathetic nerves. As a consequence of this dual supply, drooping of the upper lid, called **ptosis**, can be due to either paralysis of the oculomotor nerve or interruption of the sympathetic supply.

THE EAR

The parts of the ear are shown in Fig. 175. Sound waves are funnelled by the **auricle** (A) into the **external auditory meatus** (B). The external meatus is about one inch long; it has a skeleton of cartilage (G) in its outer third, and of the temporal bone (H) in its inner two-thirds. The skin of the external meatus contains glands which secrete wax, and hairs which normally wave the wax towards the auricle so that it does not collect in the ear.

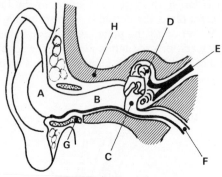

Fig. 175. Parts of the ear

The **tympanic membrane** forms the lateral wall of the middle ear (C), which is a small air-filled cavity in the temporal bone. This cavity contains the three small **ossicles** which transmit and amplify the movements of the membrane to the inner ear (D). The inner ear is an elaborate system of spaces in the bone and it is also called the **labyrinth**. It contains specialized receptors for converting sound waves to nerve impulses. Other structures in the inner ear respond to the motion of the head, as well as to its position with respect to gravity. Nerve fibres from these structures leave the inner ear through a bony canal called the internal auditory meatus (E), which opens into the cranial cavity. The Eustachian tube (F) is about $1\frac{1}{2}$ inches long and connects the middle ear with the naso-pharynx.

Middle Ear

The **middle ear** can be thought of as a room (Fig. 176) having four walls, a roof and a floor. The lateral wall, formed by the tympanic membrane, has been removed so that the medial wall is facing you, and the anterior wall is on your right. The anterior wall contains the opening (A) of the Eustachian tube, and the attachment of the **tensor tympani muscle** (B). This muscle is inserted on to the **malleus**, one of the ossicles, which is attached in its turn to the tympanic membrane (the ear drum). The tensor tympani is so called because it maintains the correct tension of the tympanic membrane.

On the posterior wall, an opening (C) is the **aditus** (doorway) to the

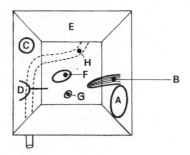

Fig. 176. Middle ear

mastoid air cells. These are hollow spaces in the mastoid process of the tympanic bone which lies just behind the ear. The aditus can transmit infection (**otitis media**) from the middle ear to the mastoid air cells. Inferior to the aditus there is a bony projection (D) called the **pyramid**; from it arises a small muscle called the **stapedius** which is attached to the smallest ossicle, the **stapes.**

The bony roof (E) of the middle-ear cavity is thin and the temporal lobe of the brain rests on its upper surface. An injury or infection of this region may, therefore, result in meningitis or a cerebral abscess.

The medial wall has two openings, or windows. The **oval window** (F) opens into the internal ear and is covered by the **foot piece** of the stapes. The **round window** (G) also opens into the internal ear and is composed of a membrane. The dotted lines (H) in Fig. 176 indicate the position of the **facial canal** which contains the facial nerve. The facial nerve leaves the cranial cavity with the auditory nerve in the internal auditory meatus, and then enters the facial canal in the medial wall of the middle ear. The canal continues down the posterior wall of the middle ear where it is closely related to the mastoid air cells. It finally leaves the temporal bone through the **stylo-mastoid foramen.** An operation on an infected mastoid may easily damage the facial nerve and cause paralysis of the facial muscles.

The **carotid canal** and the **jugular fossa** are openings in the temporal bone, closely related to the floor of the middle ear.

The function of the ossicles is shown in Fig. 177. The **malleus**, or hammer (A), is attached to the tympanic membrane (B) and its free head articulates with the **incus** (C). The incus (anvil) is articulated with the **stapes**, or stirrup (D). The footpiece of the stapes fits against the oval window (E), and the fluid in the internal ear is in contact with that part of the stapes. An annular (ring-shaped) ligament (F) attaches the footpiece of the stapes to the oval window. This ligament is lax to allow movement of the stapes.

As the tympanic membrane vibrates in response to sound waves hitting its outer surface (Fig. 177*b*) its movements are transmitted by

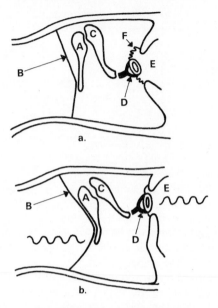

Fig. 177. Ossicles of middle ear

the ossicles to the fluid of the internal ear. The three ossicles form a lever system which diminishes the amplitude of the sound waves and increases their force on the inner ear. Excessive movement of the ossicles, due to a loud noise, is prevented by the action of the stapedius and tensor tympani muscles.

Mechanical Impairment of Hearing

Conditions which interfere with the mechanics of the transmission of sound to the internal ear are known, if permanent, as '**conductive deafness**'. Wax may form a plug in the outer ear, and so produce deafness of this type. Fluid in the middle ear will interfere with the movement of the ossicles. It is usually due to infection (otitis media) or may be caused by injury, and is another common cause of conductive deafness.

The temporary impairment of hearing experienced by air passengers during take-off or landing is due to pressure differences across the tympanic membrane. In Fig. 178 the outer surface of the tympanic membrane (A) is in contact with the atmosphere via the external auditory meatus (B). The inner surface of the membrane responds to the air pressure inside the middle ear (C). The Eustachian tube (D) allows the middle ear to communicate with the naso-pharynx, which is in contact with the atmosphere through the nose or the open mouth. Normally the pressure is the same on both sides of the ear drum, since the end of the Eustachian tube (E) is closed by the action of muscles which surround the opening. However, these muscles contract in either swallowing or yawning, and then the end of the tube gapes open.

During the ascent of an aircraft the atmospheric pressure decreases, while the pressure inside the middle ear remains the same, if the Eustachian tube is closed. The ear drum is pushed laterally by the higher pressure in the middle ear and becomes taut, which limits vibration, and so hearing is less acute. The act of swallowing or yawning will open the communication with the naso-pharynx, and equalize the pressures on both sides of the ear drum. During descent of the aeroplane, the relative pressures are reversed, and the ear drum is pushed medially into the middle ear. The treatment is the same. The

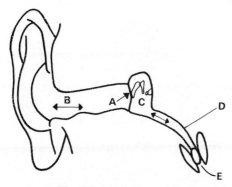

Fig. 178. Eustachian tube

Eustachian tube is frequently blocked in infections of the nasal cavity or middle ear. In such conditions, flying will give rise to earache, as the pressures cannot be equalized very easily.

Blockage of the Eustachian tube may be due to adenoids in children. **Adenoids** are masses of lymphoid tissue in the naso-pharynx. After childhood they gradually shrink and have usually disappeared in adult life. Occasionally they grow very large and block the opening of the Eustachian tube. The air in the middle ear is gradually absorbed, and as it cannot be replaced, the pressure falls, the ear drum is pushed inwards and infection is very common.

An overgrowth of bone around the oval window may involve the ligament which holds the stapes in position. In this condition, known as **otosclerosis**, deafness is due to limitation of movement of the stapes. In a high proportion of cases, hearing can be restored by surgically replacing the stapes with an artificial structure. If this is not possible, a new window may be made in the wall between the middle ear and the internal ear. This is covered by a thin flap of skin and allows the passage of sound waves to the cochlea.

Internal Ear

The **internal ear** consists of the **membranous labyrinth**, which is two connected sacs containing fluid. The larger sac is the **utricle** and it gives rise to three semicircular canals. The smaller sac is the **saccule** and it gives rise to a spiral duct, the **cochlea**. The membranous labyrinth lies in the bony labyrinth, which is hollowed out of the petrous part of the temporal bone.

The spaces of the bony labyrinth are shown in Fig. 179. The upper space (S) is for the **semicircular canals**. Only one is shown here, as the two others lie in different planes. The space V is the **vestibule** which contains the two connected sacs, the utricle (A) and the saccule (B). The remaining space (C) contains the cochlea.

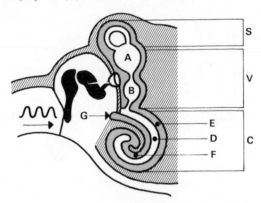

Fig. 179. Inner ear

The membranous sacs are filled with a fluid called **endolymph**. A fluid called **perilymph** fills the space between the membranous sacs and the walls of the bony labyrinth. This is indicated by the stippled area. In the cochlea, the perilymph is contained in two channels separated by the cochlear duct containing endolymph (D). The two channels are called **scalae** (Latin for stairways). The channel from the vestibule is known as the **scala vestibuli** (E) and the one leading to the middle ear, or tympanic cavity, is called the **scala tympani** (F). The scala tympani opens into the middle ear through the round window (G), which is covered by membrane to prevent leakage of the perilymph. The scala vestibuli is continuous with the scala tympani at the tip of the cochlear duct.

Sound waves enter the external auditory meatus and produce vibration of the membrane. The movement is passed through the ossicles to the oval window, where it is transmitted to the perilymph. The waves in the perilymph travel in the scala vestibuli, and may be transferred through the walls of the cochlear duct to the endolymph. Since fluid is almost incompressible, no movement can occur in the perilymph unless expansion is permitted at the other end of the channel. The waves pass to the scala tympani, and the membrane covering the round window allows temporary expansion of the perilymph. As the stapes moves to compress the perilymph at the oval window, there is a corresponding bulge of the round-window membrane into the cavity of the middle ear.

The Cochlea

This part of the inner ear is so named because it resembles the shell of a snail (Latin, *cochlea*), as it winds round a central spike of bone. A section through the cochlea is shown in Fig. 180. The scala

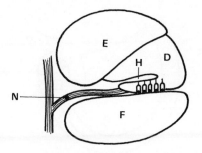

Fig. 180. Section through cochlea

vestibuli (E) and the scala tympani (F) are filled with perilymph. The **cochlear duct** (D) is filled with endolymph and contains the nerve endings that respond to sound waves. The membrane that separates the cochlear duct, or **scala media** (D), from the scala tympani is called the **basilar membrane**. Tall epithelial cells rest on the basilar membrane, and hairs projecting upwards from these cells become embedded in an overhanging roof called the **tectorial membrane** (H). This structure is called the **organ of Corti**, and it responds to pressure changes in the endolymph which produce movement of the hairs. The movement of the hairs is translated into nerve impulses that travel in the auditory nerve (N) to the temporal lobe of the cerebral cortex.

Balance

The other parts of the internal ear are concerned with the perception of the position and movements of the head, and the balance of the body.

The three semicircular canals are set in three different planes. You can visualize the arrangement by looking into the corner of a room. Two walls and the floor are set in three planes and meet at the corner. The semicircular canals meet in the utricle. The membranous duct swells as it opens into the utricle and is called the **ampulla**. One canal is shown in Fig. 181. The ampulla (A) contains the **crista**, which is a nerve ending similar to the organ of Corti. Tall epithelial cells (B) give rise to hairs (C) which are embedded in the gelatinous mass (D). The flow of endolymph in the direction of the arrow exerts tension on the hairs, and so sets up nerve impulses. The endolymph is set in motion by movements of the head. When you rotate a pan containing fluid, you can see that the apparent flow of the fluid is opposite to the movement of the pan and it continues after the pan has stopped moving. Any movement of the head will produce a current, towards the ampulla, in the endolymph of at least one canal. The impulses so set

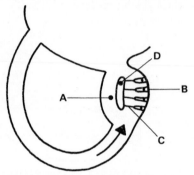

Fig. 181. Section through semicircular canal

up are interpreted as changes in the position of the head. If you spin round and stop, the endolymph will continue moving for a while and give rise to the impression that you are still spinning, but in the opposite direction. However, the rest of your body is setting up impulses that give the brain the information that the body is stationary. So, for a while, you interpret the conflicting impulses as meaning the room is spinning round you.

The utricle and saccule contain nerve ends of a similar pattern, called the **macula**. Hair cells inside the membranous sac are in contact with a gelatinous mass, which contains tiny pieces of calcium carbonate called **otoliths**. The pull of gravity on these little 'ear stones' exerts tension on the hairs and sets up nervous impulses. If the head is bent to the right, the macula of the right saccule is stimulated. If it is bent forward or backward, the macula of the utricle responds.

Impulses from all these parts are transmitted by the **vestibular branch** of the auditory nerve. The impulses reach a very complex network in the brain where impulses from the vestibular apparatus, the eyes and the sensory endings in muscles and joints, which transmit information about the position of the body, are co-ordinated. Our sense of balance and the co-ordination of bodily movements depend a great deal on information from the vestibular nerves. Any excessive stimulation of the vestibuli can produce changes in blood pressure, pulse and movements of the gastro-intestinal tract. The contradictory and confusing impulses set up when the body is inside a moving vehicle, particularly one moving in all the planes like a ship in a storm, often result in the condition known as **motion sickness**. Diseases affecting the inner ear may result in subjective feelings of giddiness, and upset the balance of the body.

THE FACE

Blood Supply

The skin of the face and scalp is very vascular, so that wounds bleed copiously but heal rapidly. Branches of the external carotid artery supply the face and scalp. The **facial artery** (A in Fig. 182) arises from the external carotid artery below the mandible, and hooks round it, where its pulsations can be felt against the bone. It travels towards the angle of the mouth and gives off branches to the upper and lower lips. It continues towards the eye to supply the more superficial structures of the cheek.

The **occipital artery** (C) runs to the back of the scalp, at first below the

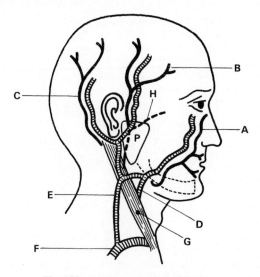

Fig. 182. Arteries and veins of the face

sterno-mastoid muscle (G) but then more superficially. The external carotid artery ends anterior to the ear, where it divides to form the superficial temporal maxillary arteries. The **superficial temporal artery** (B) can be easily seen and felt, particularly when its walls have become thick and tortuous in old age. The **internal maxillary artery** (H) lies deep to the parotid gland (P); it supplies the upper and lower jaws and the nasal cavity, and also gives off the middle meningeal artery which enters the cranial cavity to supply the dura mater and bones of the scalp.

The arteries are accompanied by veins of the same name, usually arranged as shown in Fig. 182. The **common facial vein** (D) drains into the **internal jugular vein** which lies deep to the sterno-mastoid muscle. The other veins unite to form the **external jugular vein** (E) which crosses the sterno-mastoid superficially and then runs downwards to enter the **subclavian vein** (F) about one inch above the clavicle (not shown).

The lymph nodes of the neck are arranged in chains which are closely associated with the veins. They often enlarge in infections of the mouth, pharynx, ear, nose etc., and then they can be felt.

Nerve Supply

The nerves carrying sensation from the skin of the face travel with the three divisions of the trigeminal nerve. In Fig. 183, the area indicated by A is supplied by the ophthalmic division, area B is supplied by the maxillary division, and area C is supplied by the mandibular division. The area over the angle of the jaw is supplied by spinal nerves.

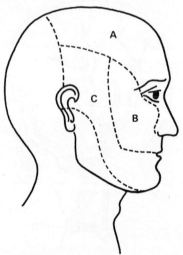

Fig. 183. Nerve supply to the skin of the face

The facial nerve is the motor supply to the muscles which control facial expression and to the buccinator (one of the muscles of mastication). All the other muscles used in mastication are supplied by the mandibular division of the trigeminal nerve.

Muscles of Mastication

There are five chief muscles used in chewing food. Two easily palpable muscles are shown in Fig. 184. The **temporalis muscle** (A) arises

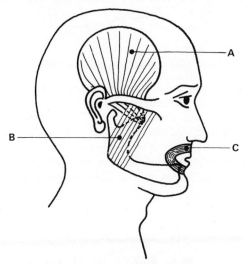

Fig. 184. Muscles of mastication (superficial)

from the **temporal fossa** and is inserted on the coronoid process of the mandible and on the anterior border of the mandibular ramus. The **masseter muscle** (B) arises from the zygomatic arch and is inserted on the lateral surface of the mandibular ramus. Both muscles contract to close the jaws. Feel them while clenching your jaws.

Two other muscles are shown in Fig. 185, where most of the ramus of the mandible has been removed along with the temporalis and masseter muscles to disclose the **infratemporal region.** The more external **lateral pterygoid muscle** (A) arises from the pterygoid process and is inserted on the neck of the mandible and the capsule of the **temporomandibular joint.** Contraction of this muscle depresses the jaw, as in opening the mouth, and also protrudes it forwards. The deeper **medial pterygoid muscle** (B) arises from the pterygoid process and is inserted on the medial surface of the angle of the mandible. It acts with the temporalis and masseter muscles to close the jaw.

The remaining muscle of mastication is the **buccinator** (C) which arises from the pterygoid process and medial surface of the mandible.

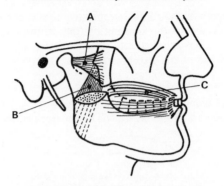

Fig. 185. Muscles of mastication (deep)

The muscle fibres run forward in the cheek to mix with the muscle fibres of the lips, the **orbicularis oris**. These muscles keep the food between the teeth and close the lips. They are also active in the expressive movements of the face and are supplied by the facial nerve.

The vessels and nerves related to the pterygoid muscles in the infra-temporal region are shown in Fig. 186, in which the ramus of the

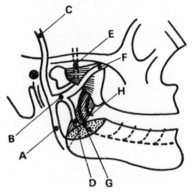

Fig. 186. Arteries and nerves of infra-temporal region

mandible and the superficial muscles have again been removed. The two pterygoid muscles are seen. The **external carotid artery** (A) ends by dividing into the superficial temporal artery (C) and the internal maxillary artery (B), which passes medial to the ramus of the mandible to give off the **inferior alveolar artery** (D) supplying the lower jaw and teeth. The **middle meningeal artery** (E) is an important branch. It passes upwards medial to the lateral pterygoid muscle to enter the cranial cavity via the foramen spinosum. The internal maxillary artery continues anteriorly (F) to enter bony channels which allow small branches to supply the nasal cavity, the hard palate, the maxillary sinus and the upper teeth.

The mandibular division of the trigeminal nerve leaves the cranial cavity through the foramen ovale; two branches can be seen emerging between the pterygoid muscles. The inferior alveolar nerve (G) accompanies the artery of the same name (D) in a bony canal in the mandible to carry sensation from the bottom teeth to the brain. It is this nerve which is 'blocked' by local anaesthetic as it enters the mandible. The remaining nerve is the **lingual** (H), which travels anteriorly to reach the tongue, and carries sensory fibres from the anterior two-thirds of the tongue.

THE TONGUE

The tongue lies in the floor of the mouth and consists of muscle covered with mucous membrane. It is shown in Fig. 187. The dorsal

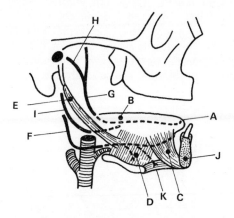

Fig. 187. Sagittal section of tongue

surface of the tongue is divided by a V-shaped groove into the **anterior part** (A), which is two-thirds of the dorsal surface and carries the taste buds, and the **posterior part** (B). The muscles of the tongue consist of two types: **intrinsic muscles**, found entirely within the tongue and which produce changes in its shape; and **extrinsic muscles**, attached to bones outside the tongue, which can move the tongue as well as changing its shape. The intrinsic muscles consist of bundles of fibres running longitudinally, vertically and transversely; they are not shown in Fig. 187.

On each side of the tongue there are three extrinsic muscles: the **genioglossus** (C) from the mandible (J); the **hyoglossus** (D) from the hyoid bone; and the **styloglossus** (E) from the styloid process. The tongue is protruded by the genioglossus, depressed by the hyoglossus and pulled into the mouth by the styloglossus.

All the muscles of the tongue receive their motor supply from the hypoglossal nerve (F), which is the twelfth cranial nerve. When one nerve is damaged, e.g. following a stroke, the tongue cannot be moved as usual: it deviates to the paralysed side. If the unconscious patient is laid on his back, the tongue falls back to obstruct the throat. This can be prevented by changing the position of the patient, or pressure on the angle of the mandible will carry the tongue forward because of the attachment of the genioglossus to the mandible.

Sensation from the anterior two-thirds of the tongue is carried in the lingual nerve (G). The impulses from the taste buds are carried in the lingual nerve, but leave it to form the **chorda tympani** (H) which joins the facial nerve to reach the brain. All sensation, including taste, from the posterior third of the tongue travels in the glossopharyngeal nerve (I).

The **geniohyoid muscle** (K) is not inserted into the tongue, but receives motor fibres from the hypoglossal nerve. The **mylohyoid muscle** (not shown in Fig. 187) forms the floor of the mouth. It stretches from the medial side of the body of the mandible to an insertion in the midline, first on the hyoid bone and then into a fibrous band which connects the right and left halves of the muscle.

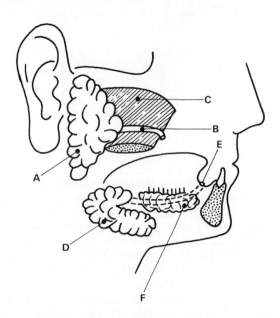

Fig. 188. Salivary glands

The Salivary Glands

There are three pairs of salivary glands; those on the right are shown in Fig. 188. The **parotid gland** (A) lies between the ramus of the mandible and the external auditory meatus. The **parotid duct** (B) carrying saliva from the gland runs forward in the cheek, and can be rolled against the anterior border of the masseter muscle (C) when the jaws are clenched. It pierces the buccinator muscle and opens into the mouth at the level of the upper second molar tooth. The facial nerve passes through the parotid gland, so an infection (**mumps**) may produce paralysis of the facial muscles.

The **submandibular gland** (D) lies at the angle of the jaw between the mandible and the posterior border of the mylohyoid muscle. The **submandibular duct** runs forward to open (E) on the floor of the mouth just below the tongue. Saliva can be seen coming from the orifices of the parotid and submandibular ducts. If you want to see them easily, suck half a lemon to promote the flow of saliva.

The **sublingual gland** (F) lies immediately below the floor of the mouth, into which it opens by a series of ducts.

THE PHARYNX

The parts of the pharynx are shown in Fig. 189. The **naso-pharynx** (A) is the posterior part of the nasal cavity and communicates behind

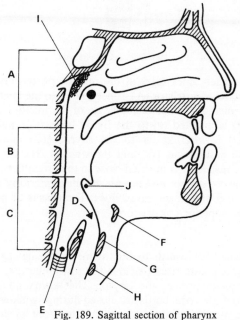

Fig. 189. Sagittal section of pharynx

the soft palate with the **oral pharynx** (B), which leads into the **laryngeal section** (C). The arrow labelled D leads into the **larynx**, which is enclosed by the **thyroid** (G) and **cricoid** (H) cartilages. The **hyoid bone** (F) is superior to the larynx. The space labelled E leads into the oesophagus. The **epiglottis** (J) lies above the larynx. The adenoids (I) are seen in the naso-pharynx just above the opening of the Eustachian tube.

The pharynx is surrounded by a muscular coat that extends from the base of the skull to the cricoid cartilage (Fig. 190). The muscles

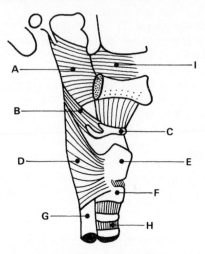

Fig. 190. Muscles of pharynx

of the right and left sides meet in the midline posteriorly, to form a tube. The **superior constrictor** (A) is inserted on to the lateral pterygoid plate and the medial surface of the angle of the mandible; the **middle constrictor** (B) is inserted on to the hyoid bone (C); and the **inferior constrictor** (D) is inserted on to the thyroid (E) and cricoid (F) cartilages. The oesophagus (G) and the trachea (H) are also shown. The superior constrictor is inserted close to the origin of the buccinator muscle (I). The constrictor muscles are all open anteriorly at the entries of the nasal, oral and laryngeal cavities. The **tonsils** lie on the medial surfaces of the superior constrictors.

Swallowing

Food is pushed backwards by the tongue squeezing against the hard palate. The soft palate rises to close the naso-pharynx, and the epiglottis descends to protect the larynx. The larynx and hyoid bone are elevated and the vocal cords are closed during swallowing. You can feel them rise by placing your fingers on the front of your throat as you swallow. If you laugh while swallowing, you nearly always feel a very painful sensation (choking) as crumbs of food or drops of liquid enter the larynx through open vocal cords.

THE LARYNX

The structures which form the framework of the larynx are seen from the lateral aspect in Fig. 191. The hyoid bone (A) is U-shaped with two processes, the **lesser horn** (B) and the **greater horn** (C), on

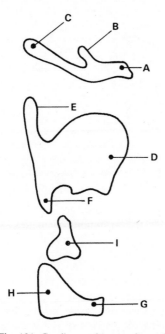

Fig. 191. Cartilages of larynx (lateral)

each side. It lies at the base of the tongue, and is attached by various muscles to the mandible, the tongue, the styloid process and the pharynx. It moves upwards with the tongue in swallowing.

The thyroid cartilage is formed by two plates (D) which meet in the midline to form the 'Adam's apple'. Posteriorly, each plate ends in a **superior horn** (E) and **inferior horn** (F). It is attached above to the hyoid bone by the **thyro-hyoid membrane**, so it moves with the hyoid in swallowing.

The cricoid cartilage is shaped like a signet ring and the **anterior arch** (G) is narrower than the **posterior wall** (H). It is attached to the thyroid cartilage by the **crico-thyroid membrane**; this membrane is free at its upper border and forms the vocal ligament inside the vocal cords.

Two pyramid-shaped cartilages, the **arytenoids** (I), sit on each side of the posterior wall of the cricoid.

In Fig. 192*a* the larynx is shown from the side after removing the right plate of the thyroid cartilage. The thyroid cartilage (A) has been cut through anteriorly, and the cricoid cartilage (B) is shown with the right arytenoid (C) sitting on its upper border. The crico-thyroid membrane (D) extends upwards from the cricoid cartilage and is inserted

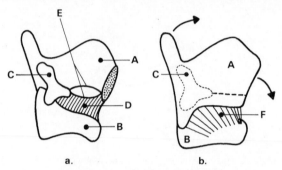

a.   b.

Fig. 192. Internal structure of larynx

into the inner surface of the thyroid cartilage at the front. The free upper border (E) is, therefore, attached anteriorly to the thyroid cartilage and posteriorly to the arytenoid cartilages. This forms the **vocal ligament**, and folds of mucosa covering the ligaments are called the **vocal cords.**

The **crico-thyroid muscle** (F) is shown in Fig. 192*b*, which is a view of the larynx from the lateral aspect. This muscle pulls the thyroid cartilage (A) forward in the direction of the arrows and so increases the length and tension of the vocal cords, as the arytenoids (C) do not move. Increasing the tension of the vocal cords raises the pitch of the voice, which is produced by forcing air between the vocal cords and so making them vibrate.

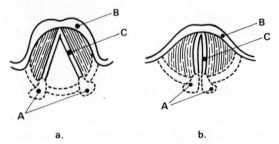

a.   b.

Fig. 193. Vocal cords

The vocal cords are seen from above in Fig. 193. The width of the slit between the cords (C) can be varied by the action of the intrinsic muscles of the larynx. In Fig. 193*a*, the arytenoids (A) are abducted, as in quiet breathing. In Fig. 193*b* they are adducted, and the passage of air through the narrow slit causes vibration of the cords, producing sounds. The epiglottis (B) is anterior to the larynx and protects the vocal cords during swallowing.

The muscles of the larynx are supplied by branches of the vagus, the tenth cranial nerve. These nerves are sometimes injured in surgical operations on the neck. If both nerves are divided, the voice is completely lost. Speech is painful and difficult in laryngitis (inflammation of the larynx) which may complicate a cold in the head.

DEEP STRUCTURES OF THE NECK

The deep structures of the neck are shown in Fig. 194. The atlas (A) and the axis (B) are the first two of the seven cervical vertebrae. The

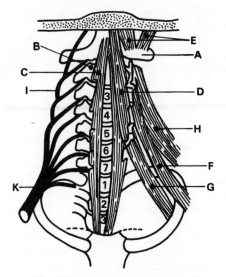

Fig. 194. Deep structures of the neck

pre-vertebral muscles run longitudinally on the anterior surfaces of the vertebrae. The **longus colli** (C) and the **longus capitis** (D), which is shown on the left only, both flex the neck. The longus capitis muscles and the short **atlanto-occipital muscles** (E) also flex the head (nodding). The three scalene muscles, also shown only on the left, originate from the transverse processes of the cervical vertebrae. The **scalenus anterior** (G) is inserted on the first rib. The **scalenus medius** (F) and **scalenus**

posterior (H) are inserted on the first and second ribs. These muscles also flex the neck. The muscles of the neck receive motor nerves from the cervical nerves (I). The roots of the **brachial plexus** (K) are seen in Fig. 194 emerging between the transverse processes of the vertebrae.

Blood-vessels

The blood vessels and nerves in the root of the neck are shown in Fig. 195. The **left subclavian artery** (A) is shown crossing over the first rib, together with the **left subclavian vein** (B) and the **brachial plexus** (H). The scalenus anterior muscle (M) is shown on the right only. The artery lies posterior to this muscle, and the vein lies anterior. The

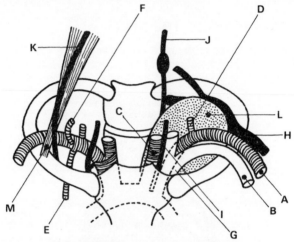

Fig. 195. Structures of the root of the neck

phrenic nerve (K) lies on this muscle as it passes down the neck from the cervical part of the spinal cord to reach the thorax and the diaphragm, which it supplies.

The **left common carotid artery** (C) starts from the arch of the aorta in the root of the neck. It passes upwards to divide into the external and internal carotid arteries. The subclavian artery gives off three branches in the neck. The **vertebral artery** (D) ascends in the transverse processes of the cervical vertebrae to reach the foramen magnum and cranial cavity. The **inferior thyroid artery** (F) sends blood to the thyroid gland. The **internal mammary artery** (E) is shown in Fig. 195 only on the right and after the removal of the right subclavian vein. It is one of the arteries supplying the breast. The subclavian vein is joined by the internal jugular vein to form the **innominate vein** (G).

The **vagus nerve** (I) is closely associated with the carotid artery and the internal jugular vein. It gives off branches to the larynx and carries parasympathetic nerves to the heart and great blood-vessels. The **cervical sympathetic trunk** (J) has three ganglia, or swellings, in its passage through the neck.

The upper part of the pleura covering the lung extends upwards into the neck. It is shown in Fig. 195 on the left only (L). The root of the neck is continuous with the superior mediastinum.

Thyroid Gland

The thyroid gland is the largest endocrine gland in the body and it is closely related to the larynx. Fig. 196 shows how the two **lobes** (E) of the gland are united by an **isthmus** (F) which lies on the anterior

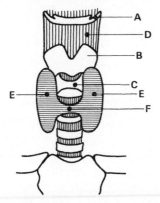

Fig. 196. Thyroid gland

aspect of the trachea. The thyroid cartilage (B) and cricoid cartilage (C) are just above the thyroid gland. The thyro-hyoid membrane (D) connects the thyroid cartilage with the hyoid bone (A).

SUPERFICIAL STRUCTURES OF THE NECK

The anterior surface of the neck is covered by muscles (Fig. 197). The largest is the **sterno-mastoid muscle** (D) which runs from the mastoid process and is inserted on to the sternum and the clavicle. The thin strap muscles are named according to their attachments also. The **thyro-hyoid** (A), the **sterno-hyoid** (B) and the **omo-hyoid** (C) are all paired, but only one of each is shown. The **mylo-hyoid** muscle (E) forms the floor of the mouth and is partly covered by the anterior belly of the **digastric muscle** (G). These muscles open the mouth by depressing the mandible (F).

If you press your jaw hard against your hand, the opposite sterno-mastoid muscle is contracted and clearly seen. This muscle (B) helps to define the **posterior triangle** of the neck (C), which is shown in Fig. 198

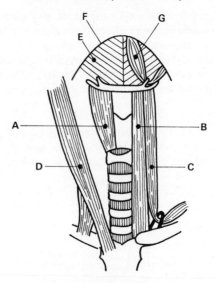

Fig. 197. Superficial structures of the neck

between the sterno-mastoid and the trapezius muscle (A). The **anterior triangle** of the neck (D) lies between the sterno-mastoid, the mandible and the midline of the neck. It is further divided by the digastric muscle (E) and the omo-hyoid muscle (F). These triangles are convenient reference points for the other structures of the neck.

In a transverse section of the neck (Fig. 199) the relationships of the structures may be reviewed. Lying on the vertebral bodies are the pre-vertebral muscles (A). Arising from the transverse processes are the scalenus anterior (B), scalenus medius and scalenus posterior (C) muscles. The roots of the cervical nerves (I) emerge between these two muscle groups. The trachea (D) is covered on the front and sides by the thyroid gland (J). The oesophagus lies posterior to the trachea.

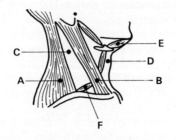

Fig. 198. Muscular triangles of the neck

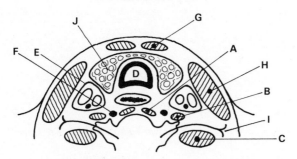

Fig. 199. Transverse section of the neck (anterior)

The common carotid artery, the internal jugular vein and the vagus nerve are all enclosed by the **carotid sheath** (E), which is fibrous and lies anterior to the sympathetic trunk (F). The strap muscles (G) are anterior to the thyroid gland. The sterno-mastoid muscle (H) is at the side of the neck.

The muscles at the back of the neck have not yet been described. The trapezius is the most superficial, and runs from the occipital prominence at the back of the skull to the clavicle, acromion and scapula. As well as extending the neck, the trapezius has important applications in the shoulder region. Other muscles, between the trapezius and the vertebrae, also take part in the movements of the head and neck.

INDEX

Page numbers set in italic type indicate diagram references.

" Switzerland - Image of a people "
        19:80 Fr.   edit alfred Vetter
pub - Benteli - Berne 1971